Pitman Research Notes in Mathematics Series

Submission of proposals for consideration
Suggestions for publication, in the form of outlines and representative samples, are invited by the Editorial Board for assessment. Intending authors should approach one of the main editors or another member of the Editorial Board, citing the relevant AMS subject classifications. Alternatively, outlines may be sent directly to the publisher's offices. Refereeing is by members of the board and other mathematical authorities in the topic concerned, throughout the world.

Preparation of accepted manuscripts
On acceptance of a proposal, the publisher will supply full instructions for the preparation of manuscripts in a form suitable for direct photo-lithographic reproduction. Specially printed grid sheets are provided and a contribution is offered by the publisher towards the cost of typing. Word processor output, subject to the publisher's approval, is also acceptable.

Illustrations should be prepared by the authors, ready for direct reproduction without further improvement. The use of hand-drawn symbols should be avoided wherever possible, in order to maintain maximum clarity of the text.

The publisher will be pleased to give any guidance necessary during the preparation of a typescript, and will be happy to answer any queries.

Important note
In order to avoid later retyping, intending authors are strongly urged not to begin final preparation of a typescript before receiving the publisher's guidelines and special paper. In this way it is hoped to preserve the uniform appearance of the series.

Longman Scientific & Technical
Longman House
Burnt Mill
Harlow, Essex, UK
(tel (0279) 26721)

Longman Scientific & Technical
Churchill Livingstone Inc.
1560 Broadway
New York, NY 10036, USA
(tel (212) 819-5453)

Titles in this series

1 Improperly posed boundary value problems
A Carasso and A P Stone
2 Lie algebras generated by finite dimensional ideals
I N Stewart
3 Bifurcation problems in nonlinear elasticity
R W Dickey
4 Partial differential equations in the complex domain
D L Colton
5 Quasilinear hyperbolic systems and waves
A Jeffrey
6 Solution of boundary value problems by the method of integral operators
D L Colton
7 Taylor expansions and catastrophes
T Poston and I N Stewart
8 Function theoretic methods in differential equations
R P Gilbert and R J Weinacht
9 Differential topology with a view to applications
D R J Chillingworth
10 Characteristic classes of foliations
H V Pittie
11 Stochastic integration and generalized martingales
A U Kussmaul
12 Zeta-functions: An introduction to algebraic geometry
A D Thomas
13 Explicit *a priori* inequalities with applications to boundary value problems
V G Sigillito
14 Nonlinear diffusion
W E Fitzgibbon III and H F Walker
15 Unsolved problems concerning lattice points
J Hammer
16 Edge-colourings of graphs
S Fiorini and R J Wilson
17 Nonlinear analysis and mechanics: Heriot-Watt Symposium Volume I
R J Knops
18 Actions of fine abelian groups
C Kosniowski
19 Closed graph theorems and webbed spaces
M De Wilde
20 Singular perturbation techniques applied to integro-differential equations
H Grabmüller
21 Retarded functional differential equations: A global point of view
S E A Mohammed
22 Multiparameter spectral theory in Hilbert space
B D Sleeman
24 Mathematical modelling techniques
R Aris
25 Singular points of smooth mappings
C G Gibson
26 Nonlinear evolution equations solvable by the spectral transform
F Calogero
27 Nonlinear analysis and mechanics: Heriot-Watt Symposium Volume II
R J Knops
28 Constructive functional analysis
D S Bridges
29 Elongational flows: Aspects of the behaviour of model elasticoviscous fluids
C J S Petrie
30 Nonlinear analysis and mechanics: Heriot-Watt Symposium Volume III
R J Knops
31 Fractional calculus and integral transforms of generalized functions
A C McBride
32 Complex manifold techniques in theoretical physics
D E Lerner and P D Sommers
33 Hilbert's third problem: scissors congruence
C-H Sah
34 Graph theory and combinatorics
R J Wilson
35 The Tricomi equation with applications to the theory of plane transonic flow
A R Manwell
36 Abstract differential equations
S D Zaidman
37 Advances in twistor theory
L P Hughston and R S Ward
38 Operator theory and functional analysis
I Erdelyi
39 Nonlinear analysis and mechanics: Heriot-Watt Symposium Volume IV
R J Knops
40 Singular systems of differential equations
S L Campbell
41 N-dimensional crystallography
R L E Schwarzenberger
42 Nonlinear partial differential equations in physical problems
D Graffi
43 Shifts and periodicity for right invertible operators
D Przeworska-Rolewicz
44 Rings with chain conditions
A W Chatters and C R Hajarnavis
45 Moduli, deformations and classifications of compact complex manifolds
D Sundararaman
46 Nonlinear problems of analysis in geometry and mechanics
M Atteia, D Bancel and I Gumowski
47 Algorithmic methods in optimal control
W A Gruver and E Sachs
48 Abstract Cauchy problems and functional differential equations
F Kappel and W Schappacher
49 Sequence spaces
W H Ruckle
50 Recent contributions to nonlinear partial differential equations
H Berestycki and H Brezis
51 Subnormal operators
J B Conway

52 Wave propagation in viscoelastic media
F Mainardi

53 Nonlinear partial differential equations and their applications: Collège de France Seminar. Volume I
H Brezis and J L Lions

54 Geometry of Coxeter groups
H Hiller

55 Cusps of Gauss mappings
T Banchoff, T Gaffney and C McCrory

56 An approach to algebraic K-theory
A J Berrick

57 Convex analysis and optimization
J-P Aubin and R B Vintner

58 Convex analysis with applications in the differentiation of convex functions
J R Giles

59 Weak and variational methods for moving boundary problems
C M Elliott and J R Ockendon

60 Nonlinear partial differential equations and their applications: Collège de France Seminar. Volume II
H Brezis and J L Lions

61 Singular systems of differential equations II
S L Campbell

62 Rates of convergence in the central limit theorem
Peter Hall

63 Solution of differential equations by means of one-parameter groups
J M Hill

64 Hankel operators on Hilbert space
S C Power

65 Schrödinger-type operators with continuous spectra
M S P Eastham and H Kalf

66 Recent applications of generalized inverses
S L Campbell

67 Riesz and Fredholm theory in Banach algebra
B A Barnes, G J Murphy, M R F Smyth and T T West

68 Evolution equations and their applications
F Kappel and W Schappacher

69 Generalized solutions of Hamilton-Jacobi equations
P L Lions

70 Nonlinear partial differential equations and their applications: Collège de France Seminar. Volume III
H Brezis and J L Lions

71 Spectral theory and wave operators for the Schrödinger equation
A M Berthier

72 Approximation of Hilbert space operators I
D A Herrero

73 Vector valued Nevanlinna Theory
H J W Ziegler

74 Instability, nonexistence and weighted energy methods in fluid dynamics and related theories
B Straughan

75 Local bifurcation and symmetry
A Vanderbauwhede

76 Clifford analysis
F Brackx, R Delanghe and F Sommen

77 Nonlinear equivalence, reduction of PDEs to ODEs and fast convergent numerical methods
E E Rosinger

78 Free boundary problems, theory and applications. Volume I
A Fasano and M Primicerio

79 Free boundary problems, theory and applications. Volume II
A Fasano and M Primicerio

80 Symplectic geometry
A Crumeyrolle and J Grifone

81 An algorithmic analysis of a communication model with retransmission of flawed messages
D M Lucantoni

82 Geometric games and their applications
W H Ruckle

83 Additive groups of rings
S Feigelstock

84 Nonlinear partial differential equations and their applications: Collège de France Seminar. Volume IV
H Brezis and J L Lions

85 Multiplicative functionals on topological algebras
T Husain

86 Hamilton-Jacobi equations in Hilbert spaces
V Barbu and G Da Prato

87 Harmonic maps with symmetry, harmonic morphisms and deformations of metrics
P Baird

88 Similarity solutions of nonlinear partial differential equations
L Dresner

89 Contributions to nonlinear partial differential equations
C Bardos, A Damlamian, J I Díaz and J Hernández

90 Banach and Hilbert spaces of vector-valued functions
J Burbea and P Masani

91 Control and observation of neutral systems
D Salamon

92 Banach bundles, Banach modules and automorphisms of C*-algebras
M J Dupré and R M Gillette

93 Nonlinear partial differential equations and their applications: Collège de France Seminar. Volume V
H Brezis and J L Lions

94 Computer algebra in applied mathematics: an introduction to MACSYMA
R H Rand

95 Advances in nonlinear waves. Volume I
L Debnath

96 FC-groups
M J Tomkinson

97 Topics in relaxation and ellipsoidal methods
M Akgül

98 Analogue of the group algebra for topological semigroups
H Dzinotyiweyi

99 Stochastic functional differential equations
S E A Mohammed

100 Optimal control of variational inequalities
V Barbu
101 Partial differential equations and dynamical systems
W E Fitzgibbon III
102 Approximation of Hilbert space operators. Volume II
C Apostol, L A Fialkow, D A Herrero and D Voiculescu
103 Nondiscrete induction and iterative processes
V Ptak and F-A Potra
104 Analytic functions – growth aspects
O P Juneja and G P Kapoor
105 Theory of Tikhonov regularization for Fredholm equations of the first kind
C W Groetsch
106 Nonlinear partial differential equations and free boundaries. Volume I
J I Díaz
107 Tight and taut immersions of manifolds
T E Cecil and P J Ryan
108 A layering method for viscous, incompressible L_p flows occupying R^n
A Douglis and E B Fabes
109 Nonlinear partial differential equations and their applications: Collège de France Seminar. Volume VI
H Brezis and J L Lions
110 Finite generalized quadrangles
S E Payne and J A Thas
111 Advances in nonlinear waves. Volume II
L Debnath
112 Topics in several complex variables
E Ramírez de Arellano and D Sundararaman
113 Differential equations, flow invariance and applications
N H Pavel
114 Geometrical combinatorics
F C Holroyd and R J Wilson
115 Generators of strongly continuous semigroups
J A van Casteren
116 Growth of algebras and Gelfand–Kirillov dimension
G R Krause and T H Lenagan
117 Theory of bases and cones
P K Kamthan and M Gupta
118 Linear groups and permutations
A R Camina and E A Whelan
119 General Wiener–Hopf factorization methods
F-O Speck
120 Free boundary problems: applications and theory, Volume III
A Bossavit, A Damlamian and M Fremond
121 Free boundary problems: applications and theory, Volume IV
A Bossavit, A Damlamian and M Fremond
122 Nonlinear partial differential equations and their applications: Collège de France Seminar. Volume VII
H Brezis and J L Lions
123 Geometric methods in operator algebras
H Araki and E G Effros
124 Infinite dimensional analysis–stochastic processes
S Albeverio
125 Ennio de Giorgi Colloquium
P Krée
126 Almost-periodic functions in abstract spaces
S Zaidman
127 Nonlinear variational problems
A Marino, L Modica, S Spagnolo and M Degiovanni
128 Second-order systems of partial differential equations in the plane
L K Hua, W Lin and C-Q Wu
129 Asymptotics of high-order ordinary differential equations
R B Paris and A D Wood
130 Stochastic differential equations
R Wu
131 Differential geometry
L A Cordero
132 Nonlinear differential equations
J K Hale and P Martinez-Amores
133 Approximation theory and applications
S P Singh
134 Near-rings and their links with groups
J D P Meldrum
135 Estimating eigenvalues with *a posteriori/a priori* inequalities
J R Kuttler and V G Sigillito
136 Regular semigroups as extensions
F J Pastijn and M Petrich
137 Representations of rank one Lie groups
D H Collingwood
138 Fractional calculus
G F Roach and A C McBride
139 Hamilton's principle in continuum mechanics
A Bedford
140 Numerical analysis
D F Griffiths and G A Watson
141 Semigroups, theory and applications. Volume I
H Brezis, M G Crandall and F Kappel
142 Distribution theorems of L-functions
D Joyner
143 Recent developments in structured continua
D De Kee and P Kaloni
144 Functional analysis and two-point differential operators
J Locker
145 Numerical methods for partial differential equations
S I Hariharan and T H Moulden
146 Completely bounded maps and dilations
V I Paulsen
147 Harmonic analysis on the Heisenberg nilpotent Lie group
W Schempp
148 Contributions to modern calculus of variations
L Cesari
149 Nonlinear parabolic equations: qualitative properties of solutions
L Boccardo and A Tesei
150 From local times to global geometry, control and physics
K D Elworthy

151 A stochastic maximum principle for optimal control of diffusions
U G Haussmann

152 Semigroups, theory and applications. Volume II
H Brezis, M G Crandall and F Kappel

153 A general theory of integration in function spaces
P Muldowney

154 Oakland Conference on partial differential equations and applied mathematics
L R Bragg and J W Dettman

155 Contributions to nonlinear partial differential equations. Volume II
J I Díaz and P L Lions

156 Semigroups of linear operators: an introduction
A C McBride

157 Ordinary and partial differential equations
B D Sleeman and R J Jarvis

158 Hyperbolic equations
F Colombini and M K V Murthy

159 Linear topologies on a ring: an overview
J S Golan

160 Dynamical systems and bifurcation theory
M I Camacho, M J Pacifico and F Takens

161 Branched coverings and algebraic functions
M Namba

162 Perturbation bounds for matrix eigenvalues
R Bhatia

163 Defect minimization in operator equations: theory and applications
R Reemtsen

164 Multidimensional Brownian excursions and potential theory
K Burdzy

165 Viscosity solutions and optimal control
R J Elliott

Makoto Namba

Tôhoku University

Branched coverings and algebraic functions

Longman Scientific & Technical

Copublished in the United States with
John Wiley & Sons, Inc., New York

Longman Scientific & Technical
Longman Group UK Limited
Longman House, Burnt Mill, Harlow
Essex CM20 2JE, England
and Associated Companies throughout the world.

Copublished in the United States with
John Wiley & Sons, Inc., 605 Third Avenue, New York, NY 10158

First published 1987

AMS Subject Classifications: 14–XX, 32–XX

ISSN 0269-3674

British Library Cataloguing in Publication Data
Namba, Makoto
Branched coverings and algebraic functions.
—(Pitman research notes in mathematics
series; ISSN 0269-3674; 161).
1. Geometry, Profective 2. Geometry,
Algebraic
I. Title
516′.5 QA471
ISBN 0-582-01371-2

Library of Congress Cataloging-in-Publication Data
Namba, Makoto, 1943–
Branched coverings and algebraic functions.
(Pitman research notes in mathematics series ; 161)
Bibliography: p.
Includes index.
1. Complex manifolds. 2. Functions of several
complex variables. 3. Covering spaces (Topology)
I. Title. II. Series.
QA331.N26 1987 516.3′6 87-16945
ISBN 0–470–20949–6 (USA only)

Printed and bound in Great Britain by
Biddles Ltd, Guildford and King's Lynn

To my teacher

Professor Masatake Kuranishi

Contents

Preface

In this book, I present a theory of branched coverings of projective manifolds in connection with the theory of algebraic functions of several complex variables.

This book has originated from a series of lectures and seminar talks for graduate students given at Tokyo Institute of Technology, Tokyo Metropolitan University and Kyushu University.

I would like to express my deep gratitude to Professors M. Oka, J. Noguchi, N. Sasakura, Mi. Kato and M. Kita for many valuable suggestions and stimulating discussions.

I also thank Mr. Konno who read the first draft and corrected mistakes.

Makoto Namba

Introduction

- Looking up at stars twinkling
in the night sky -

The purpose of this book is to present a theory of finite branched coverings of projective complex manifolds in connection with the theory of algebraic functions of several complex variables.

For a given projective manifold M, a *finite branched covering of* M is, by definition, an irreducible normal complex space X together with a surjective proper finite holomorphic mapping $\pi : X \longrightarrow M$. The *ramification locus* R_π *of* π is the set of points $x \in X$ such that $\pi^* : \mathcal{O}_{M,\pi(x)} \longrightarrow \mathcal{O}_{X,x}$ is not isomorphic. ($\mathcal{O}_{X,x}$ is the local ring of germs of holomorphic functions around x.) The *branch locus* B_π *of* π is the image $\pi(R_\pi)$ under π. Then R_π and B_π are hypersurfaces of X and M, respectively. To any irreducible component D' of $\pi^{-1}(B_\pi)$, a positive integer $e_{D'}$ is associated and is called the *ramification index of* π *at* D'. The *covering transformation group* G_π *of* π is the group of all automorphisms (holomorphic isomorphisms onto itself) $\varphi \in \mathrm{Aut}(X)$ such that $\pi\varphi = \pi$. π is said to be *Galois* if G_π acts transitively on every fibre of π. π is said to be *abelian* if π is Galois and G_π is an abelian group.

Suppose that irreducible hypersurfaces $D_1,\ldots,D_s$ of M are given. Put $B = D_1 \cup\ldots\cup D_s$. Suppose also that positive integers $e_1,\ldots,e_s$ are given. Consider the positive divisor $D = e_1D_1 + \ldots + e_sD_s$ on M. A finite branched covering $\pi : X \longrightarrow M$ is said to *branch at* D (resp. *at most at* D) if $B_\pi = B$ (resp. $B_\pi \subset B$) and, for every j and for any irreducible component D' of $\pi^{-1}(D_j)$, the ramification index at D' is e_j (resp. divides e_j).

In this book, we discuss the following two problems:

PROBLEM 1. Give a necessary and sufficient condition on (M,D) for the existence of a finite Galois (resp. abelian) covering $\pi : X \longrightarrow M$ which branches at D.

PROBLEM 2. Describe the set of all (isomorphism classes of) finite Galois (resp. abelian) coverings $\pi : X \longrightarrow M$ which branch at most at D.

The first problem may be called *Fenchel's problem.* Fenchel's original problem is for compact Riemann surfaces M and was answered by Bundgaard-Nielsen [4] and Fox [14] (cf. section 1.2).

It is noted that (isomorphism classes of) finite Galois branched coverings $\pi : X \longrightarrow M$ and (isomorphism classes of) finite Galois extensions $K/\mathbb{C}(M)$ are in one-to-one correspondence, where $\mathbb{C}(M)$ is the field of meromorphic functions on M;

$$\pi \longmapsto \mathbb{C}(X)/\mathbb{C}(M) \quad \text{and} \quad K/\mathbb{C}(M) \longmapsto K\text{-normalization of } M.$$

Hence the second problem is class field theoretic. Historically, Hilbert conjectured the (absolute) class field theory as an analogue of the theory of covering spaces of compact Riemann surfaces. The class field theory by Takagi and Artin is one of the deepest and the most beautiful mathematical theories, but it is said that the non-abelian class field theory is not completely constructed yet. The famous paper [64] by Weil seems to construct a model of the non-abelian theory using fields of algebraic functions of one variable. In this book, we follow Weil's paper by rewriting (parts of) it for fields of algebraic functions of several variables (cf. chapter 3).

Now we introduce each chapter. In chapter 1, we discuss branched coverings of complex manifolds. We take *not necessarily finite* coverings into consideration and introduce the notion of *maximal coverings which branch at* D, which play a rôle in our theory. We give conditions for the existence of maximal coverings. Of course, we are mainly interested in finite branched coverings. We give many examples of finite branched coverings. In particular, we explain one of the interesting examples by Hirzebruch [30]. In section 1.5, we give a generalization of Kato's theorem on Fenchel's problem.

In chapter 2, we give answers to the above problems for finite abelian coverings, by using the theory of currents by de Rham-Kodaira [12], (cf. Theorems 2.3.18 and 2.3.20). This chapter can be regarded as a higher dimensional generalization (with ramification) of Iwasawa [35, chapter 5]. At the end of this chapter, we describe the set of all (isomorphism classes of) finite abelian branched coverings of M, using the notion of rational divisor classes (cf. Theorem 2.3.30).

In chapter 3, we give answers to the above problems for finite Galois branched coverings (cf. Theorems 3.5.3 and 3.5.5), following Weil [64], Tōyama [62] and Grothendieck [23]. At the end of this chapter, we give the following theorem (cf. Theorem 5.3.7):

THEOREM. There is a one-to-one correspondence $\pi \longrightarrow S = S(\pi)$ between the set of all isomorphism classes of finite Galois branched coverings $\pi : X \longrightarrow M$ and the set of all finite modules S of UFGV(M), where UFGV(M) is the Tannaka system of all unitary flat generalized vector bundles on M. The correspondence satisfies (1) $G_\pi \simeq G(S(\pi))$ (≒ the representation group of $S(\pi)$) and (2) $\pi_1 \leq \pi_2$ if and only if $S(\pi_1) \subset S(\pi_2)$.

This theorem says also that the set of all (isomorphism classes of) finite Galois extensions $K/\mathbb{C}(M)$ can be controlled by UFGV(M), (cf. Theorem 5.3.7').

This theorem, however, is unsatisfactory, because we cannot give a characterization of unitary flat generalized vector bundles using *languages which come from the inside of* M, according to the spirit of class field theory. (As for usual unitary flat vector bundles, Mehta-Ramanathan [40] recently gave a characterization using the language of stability.) It seems that many things are unsettled in this direction.

It should be noted that the theory of automorphic functions is highly developed today. This theory starts from a given bounded domain Ω in $\mathbb{C}^n$ and a given discrete subgroup G of $\mathrm{Aut}(\Omega)$. On the contrary, our theory starts from a given projective manifold M. The theory of automorphic functions is as if looking down at the earth from the sky. Our theory is as if looking up at the sky from the earth.

Our theory is far from being satisfactory. It would be deeply pleasing if an interested reader could develop the theory further.

1 Branched coverings of complex manifolds

1.1 Definition of branched coverings of complex manifolds

We begin our discussion with the definition of branched coverings of complex manifolds:

DEFINITION 1.1.1 Let M be an n-dimensional (connected) complex manifold. A *branched covering of* M is an n-dimensional irreducible normal complex space X together with a surjective holomorphic mapping $\pi : X \longrightarrow M$ such that

(1) every fibre of π is discrete in X,

(2) $R_\pi = \{q \in X \mid \pi^* : \mathcal{O}_{\pi(q),M} \longrightarrow \mathcal{O}_{q,X}$ is not isomorphic$\}$,

the *ramification locus*, and $B_\pi = \pi(R_\pi)$, the *branch locus*, are hypersurfaces of X and M, respectively,

(3) $\pi : X - \pi^{-1}(B_\pi) \longrightarrow M-B_\pi$ is an unramified (i.e. topological) covering and

(4) for every point $p \in M$, there is a connected open neighbourhood W of p in M such that, for every connected component U of $\pi^{-1}(W)$,
(i) $\pi^{-1}(p) \cap U = \{q\}$ is one point and (ii) $\pi_U = \pi|U : U \longrightarrow W$ is surjective and proper (hence finite).

A branched covering $\pi : X \longrightarrow M$ is said to be *of non-singular type* (resp. *of simple normally crossing type*) if X is a complex manifold and R_π and B_π are non-singular (resp. simple normally crossing).

A branched covering $\pi : X \longrightarrow M$ is said to be *finite* if π is a finite map. The covering degree of $\pi : X-\pi^{-1}(B_\pi) \longrightarrow M - B_\pi$ is called the *degree of* π and is denoted by $\deg \pi$.

EXAMPLE 1.1.2 Let $\pi : X \longrightarrow M$ be any surjective proper *finite* holomorphic map. Then π is a finite branched covering of M. In fact, (1), (3) and (4) of Definition 1.1.1 are clearly satisfied in this case and (2) follows from the *purity of ramification locus* (see Fischer [13]). Let $\mu_o : X_o \longrightarrow X$ be the universal covering space of X. Then $\pi\mu_o : X_o \longrightarrow M$ is a (not necessarily finite) branched covering of M.

EXAMPLE 1.1.3 Let X be an n-dimensional irreducible normal complex space and G be a subgroup of $\mathrm{Aut}(X)$ acting on X properly discontinuously such that the quotient space $X/G = M$ is a complex manifold. Then the projection $\pi : X \longrightarrow M$ is a branched covering. (This follows again from the purity of ramification locus.)

REMARK 1.1.4 (1) The theory of automorphic functions starts from a bounded domain D, a discrete subgroup G of $\mathrm{Aut}(D)$ and the projection $\pi : D \longrightarrow D/G$. On the contrary, our theory starts from a complex manifold M and branched coverings $\pi : X \longrightarrow M$ of M. (2) In Definition 1.1.1, we do not avoid the case of R_π and B_π being empty. Though $\pi : X \longrightarrow M$ in this case should be called an unramified (or unbranched) covering, we still call it a branched covering by abuse of language. (3) We are mainly interested in finite branched coverings, but we must also take infinite branched coverings into consideration for a theoretical reason. (4) For a branched covering $\pi : X \longrightarrow M$, π is proper if and only if π is a finite map. (5) For a point $p \in M$, if W and U have the property in (4) of Definition 1.1.1, then any connected open neighbourhood W' of p such that $W' \subset W$ and $U' = U \cap \pi^{-1}(W')$ has the same property.

Two branched coverings $\pi : X \longrightarrow M$ and $\pi' : X' \longrightarrow M$ of M are said to be *equivalent* if there is a biholomorphic map $\varphi : X \longrightarrow X'$ such that $\pi = \pi'\varphi$. This is an equivalence relation. φ is called an *isomorphism of* π *to* π'.

A *covering transformation of a branched covering* $\pi : X \longrightarrow M$ is, by definition, an automorphism σ of $\pi : X \longrightarrow M$. The set of all covering transformations of π forms a group G_π in a natural way and is called the *covering transformation group*.

A branched covering $\pi : X \longrightarrow M$ is called a *Galois covering* if G_π acts transitively on every fibre of π. $\pi : X \longrightarrow M$ is called an *abelian* (resp. *cyclic*) *covering* if π is a Galois covering and G_π is an abelian (resp. cyclic) group.

PROPOSITION 1.1.5 Two branched coverings $\pi : X \longrightarrow M$ and $\pi' : X' \longrightarrow M$ with $B_\pi = B_{\pi'}$ are equivalent if and only if $\pi_1 : X - \pi^{-1}(B_\pi) \longrightarrow M - B_\pi$ and $\pi_1' : X' - \pi'^{-1}(B_{\pi'}) \longrightarrow M - B_{\pi'}$ are equivalent. (π_1 and π_1' are restrictions of π and π', respectively.)

PROOF. An isomorphism φ of π to π' naturally induces an isomorphism of $\pi_1 : X - \pi^{-1}(B_\pi) \longrightarrow M - B_\pi$ to $\pi_1' : X' - \pi'^{-1}(B_{\pi'}) \longrightarrow M - B_{\pi'}$.

Conversely, let $\varphi_1 : X - \pi^{-1}(B_\pi) \longrightarrow X' - \pi'^{-1}(B_{\pi'})$ be an isomorphism of π_1 to π_1'. We want to extend φ_1 to an isomorphism of π to π'. For any point q of X, let W be a connected open neighbourhood of $p = \pi(q)$ in M which satisfies the condition of (4) of Definition 1.1.1 with respect to both π and π'. Let U be the connected component of $\pi^{-1}(W)$ such that $q \in U$. Put $U_1 = U - \pi^{-1}(B_\pi)$. Since $\varphi_1(U_1)$ is connected and is contained in $\pi'^{-1}(W)$, there is a unique connected component U' of $\pi'^{-1}(W)$ such that $\varphi_1(U_1) \subset U'$. Then $\pi'^{-1}(p) \cap U' = \{q'\}$ is one point, $\pi' : U' \longrightarrow W$ is surjective and proper and $U' - \pi'^{-1}(B_{\pi'}) = \varphi_1(U_1)$. We define $\varphi : X \longrightarrow X'$ by putting

$$\varphi(q) = q'.$$

Then it can be easily checked that φ is a homeomorphism of X onto X' such that $\pi'\varphi = \pi$ and $\varphi|(X-B_\pi) = \varphi_1$.

By *Riemann's extension theorem*, φ is holomorphic on X - Sing$X$. Since X is normal, φ is holomorphic on X (see Fischer [13]). In a similar way, φ^{-1} is holomorphic on X'. Hence φ is an isomorphism of π to π'. (The correspondence $\varphi \longrightarrow \varphi_1$ is one-to-one.) □

More generally, a *morphism* φ of $\pi : X \longrightarrow M$ to $\pi' : X' \longrightarrow M$ is, by definition, a holomorphic map φ of X onto X' such that $\pi = \pi'\varphi$:

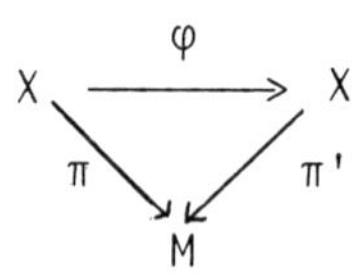

For points $p \in M$, $q \in X$ and $q' \in X'$ such that $\pi(q) = p$ and $\varphi(q) = q'$, the diagram

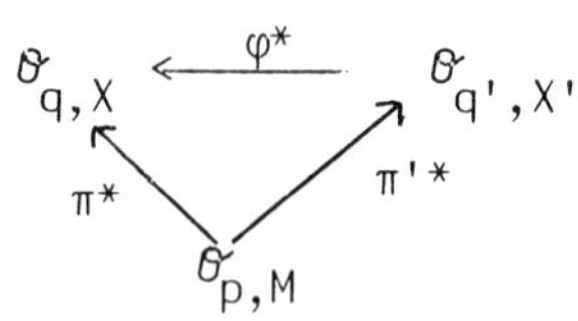

commutes. Since every fibre of π, π' and φ is discrete, π^*, π'^* and φ^* are injective. Hence π^* is isomorphic if and only if φ^* and π'^* are isomorphic. Thus

$$R_\pi = \varphi^{-1}(R_{\pi'}) \cup R_\varphi \quad \text{and} \quad B_{\pi'} \subset B_\pi.$$

A similar proof to that of Proposition 1.1.5 shows

PROPOSITION 1.1.6 Let $\pi : X \longrightarrow M$ and $\pi' : X' \longrightarrow M$ be branched coverings such that $B_{\pi'} \subset B_\pi$. Then there is a natural one-to-one correspondence between {morphisms φ of π to π'} and {morphisms φ_1 of $\pi_1 : X - \pi^{-1}(B_\pi) \longrightarrow M - B_\pi$ to $\pi_1' : X' - \pi'^{-1}(B_\pi) \longrightarrow M - B_\pi$}.

THEOREM 1.1.7 Let $\pi : X \longrightarrow M$ be a branched covering. Then (1) G_π can be naturally identified with G_{π_1}, where $\pi_1 : X - \pi^{-1}(B_\pi) \longrightarrow M - B_\pi$ is the restriction of π. (2) $\pi : X \longrightarrow M$ is Galois (resp. abelian, resp. cyclic) if and only if $\pi_1 : X - \pi^{-1}(B_\pi) \longrightarrow M - B_\pi$ is a Galois (resp. abelian, resp. cyclic) covering. (3) G_π acts on X properly discontinuously and (4) If $\pi : X \longrightarrow M$ is a Galois covering, then the complex space X/G_π is canonically biholomorphic to M.

PROOF. (1) follows from Proposition 1.1.5.

(2) If $\pi : X \longrightarrow M$ is a Galois covering, then $\pi_1 : X - \pi^{-1}(B_\pi) \longrightarrow M - B_\pi$ is clearly a Galois covering. Conversely, suppose that π_1 is a Galois covering. For any point $p \in B_\pi$, take points q and q' in $\pi^{-1}(p)$. We show that there is $\sigma \in G_\pi$ such that $\sigma(q) = q'$. Let W be a connected open neighbourhood of p which satisfies the condition of (4) of Definition 1.1.1. Let U and U' be the connected components of $\pi^{-1}(W)$ such that $q \in U$ and $q' \in U'$. Take $r \in U_1 = U - \pi^{-1}(B_\pi)$ and $r' \in U_1' = U' - \pi^{-1}(B_\pi)$ such that $\pi(r) = \pi(r')$. By the assumption, there is a covering transformation σ_1 of π_1 such that $\sigma_1(r) = r'$. It is then clear that $\sigma_1(U_1) = U_1'$. By (1), σ_1 can be uniquely extended to a covering transformation σ of π. Since $\sigma(U) = U'$, and $\pi(\sigma(q)) = p = \pi(q)$, we have $\sigma(q) = q'$.

(3) Let K_1 and K_2 be compact sets in X. We show that the set

$$G(K_1,K_2) = \{\sigma \in G_\pi \mid \sigma(K_1) \cap K_2 \text{ is non-empty}\}$$

is a finite set. Assume the converse. Then there are squences $\{\sigma_\nu\}$, $\{q_\nu\}$ and $\{q_\nu'\}$ of points of G_π, K_1 and K_2, respectively, such that $\sigma_\nu \neq \sigma_\mu$ for $\nu \neq \mu$ and $\sigma_\nu(q_\nu) = q_\nu'$ for all ν. Since K_1 and K_2 are compact, we may assume that $\{q_\nu\}$ (resp. $\{q_\nu'\}$) converges to q (resp. q'). Since $\pi(q_\nu) = \pi(q_\nu')$, for all ν, we have $\pi(q) = \pi(q') = p$. Let W, U and U' be connected open neighbourhoods of p, q, q', respectively, as in the proof of (2). We may assume that $q_\nu \in U$ and $q_\nu' \in U'$ for $\nu = 1, 2, \ldots$. Since $\sigma_\nu(U)$ is a connected component of $\pi^{-1}(W)$ and $\sigma_\nu(q_\nu) = q_\nu' \in U'$, we have $\sigma_\nu(U) = U'$ for $\nu = 1, 2, \ldots$. Hence $\sigma_1^{-1}\sigma_\nu : U \longrightarrow U$ for $\nu = 1, 2, \ldots$, induce mutually distinct covering transformations of the finite unramified covering

$\pi_U : U - \pi^{-1}(B_\pi) \longrightarrow W - B_\pi$, a contradiction. Hence $G(K_1,K_2)$ is a finite set. Thus G_π acts on X properly discontinuously.

(4) If $\pi : X \longrightarrow M$ is a Galois covering, then there is a natural bijection $\rho : X/G_\pi \longrightarrow M$. We first show that this is a homeomorphism. It suffices to show that if Z is a subset of M such that $\pi^{-1}(Z)$ is open in X, then Z is open in M. Take a sequence $\{p_\nu\}$ of points in $M - Z$ which converges to $p \in M$. Take a point $q \in X$ such that $\pi(q) = p$. Let W and U be connected open neighbourhoods of p and q, respectively, which satisfy the condition of (4) of Definition 1.1.1. We may assume that $p_\nu \in W$ for $\nu = 1, 2, \ldots$. Take points $q_\nu \in U$ such that $\pi(q_\nu) = p_\nu$. Then $\{q_\nu\}$ converges to q. Since q_ν are points of the closed set $X = \pi^{-1}(Z)$, q is also a point of $X - \pi^{-1}(Z)$. Hence $p = \pi(q) \in M - Z$. Thus Z is open in M.

Next we show that $\rho : X/G_\pi \longrightarrow M$ is biholomorphic. Let W be an open set of M and f be a holomorphic function on $\pi^{-1}(W)$ such that $f\sigma = f$ for all $\sigma \in G$. Since $\pi : X \longrightarrow M$ is a Galois covering, f induces a function g on W such that $f = g\pi$. The function g is clearly continuous on W and is holomorphic on $W - B_\pi$. By Riemann's extension theorem, g is holomorphic on W.

This means that $\rho : X/G_\pi \longrightarrow M$ is biholomorphic. □

<u>THEOREM 1.1.8</u> Let $\pi : X \longrightarrow M$ be a branched covering of M. Let p be a non-singular point of B_π. Then every point q of $\pi^{-1}(p)$ is a non-singular point of both X and $\pi^{-1}(B_\pi)$. Moreover, if W is a sufficiently small connected open neighbourhood of p with a coordinate system $(w_1,\ldots,w_n)$ such that $p = (0,\ldots,0)$ and $B_\pi \cap W = \{(w_1,\ldots,w_n) \in W \mid w_n = 0\}$, then there are a coordinate system $(z_1,\ldots,z_n)$ in the connected component U of $\pi^{-1}(W)$ with $q \in U$ and a positive integer e such that $q = (0,\ldots,0)$ and

$$\pi_U : (z_1,\ldots,z_n) \in U \longmapsto (w_1,\ldots,w_n) = (z_1,\ldots,z_{n-1},z_n^e) \in W.$$

($q \in R_\pi$ if and only if $e \geq 2$.)

<u>PROOF</u>. We may assume that W satisfies the condition of (4) of Definition 1.1.1. Moreover, we may identify W and $W - B_\pi$ with $\mathbb{D}\times\ldots\times\mathbb{D}$ (n-times) and $\mathbb{D}\times\ldots\times\mathbb{D}\times\mathbb{D}^*$, respectively, where

$$\mathbb{D} = \{w \in \mathbb{C} \mid |w| < 1\} \quad \text{(the unit disc)} \quad \text{and}$$

$$\mathbb{D}^* = \mathbb{D} - \{0\}.$$

Then, for a fixed point $r \in W - B_\pi$, the fundamental Group $\pi_1(W - B_\pi,r)$ is

the free abelian group $\alpha^{\mathbb{Z}}$ generated by a circle α around p, passing through r and oriented in a counterclockwise sense :
$\alpha = \{(r_1,\ldots,r_{n-1},r_n e^{i\vartheta}) \mid 0 \leq \vartheta \leq 2\pi\}$, where $r = (r_1,\ldots,r_n)$, $(r_n \neq 0)$.

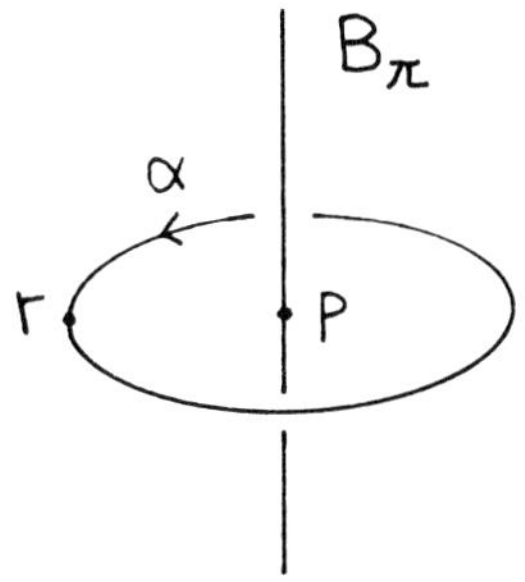

Note that $\pi : U - \pi^{-1}(B_\pi) \longrightarrow W - B_\pi$ is a finite unramified covering. Hence $U - \pi^{-1}(B_\pi)$ can be identified with $\mathbf{D} \times \ldots \times \mathbf{D} \times \mathbf{D}^*$ and hence π can be given by

$$\pi : (z_1,\ldots,z_n) \in U - \pi^{-1}(B_\pi) \longmapsto (w_1,\ldots,w_n) = (z_1,\ldots,z_{n-1},z_n^e) \in W - B_\pi$$

for some positive integer e. We put $U' = \mathbf{D} \times \ldots \times \mathbf{D}$ (n-times) $(\supset U - \pi^{-1}(B_\pi))$ and define a map $\pi' : U' \longrightarrow W$ by

$$\pi' : (z_1,\ldots,z_n) \in U' \longmapsto (w_1,\ldots,w_n) = (z_1,\ldots,z_{n-1},z_n^e) \in W.$$

Then $\pi' : U' \longrightarrow W$ is a branched covering of non-singular type. By Proposition 1.1.5, $\pi : U \longrightarrow W$ is equivalent to $\pi' : U' \longrightarrow W$. This completes the proof of the theorem. □

We note that Roan [51] gave a direct proof of the last part of the theorem.

<u>COROLLARY 1.1.9</u> Let $\pi : X \longrightarrow M$ be a branched covering such that the branch locus B_π is non-singular. Then X and $\pi^{-1}(B_\pi)$ are non-singular and hence $\pi : X \longrightarrow M$ is of non-singular type.

<u>COROLLARY 1.1.10</u> Let $\pi : X \longrightarrow M$ be a branched covering. Then $\mathrm{Sing}X \subset \pi^{-1}(\mathrm{Sing}B_\pi)$, $\mathrm{Sing}\pi^{-1}(B_\pi) \subset \pi^{-1}(\mathrm{Sing}B_\pi)$ and

$$\pi : X - \pi^{-1}(\mathrm{Sing}B_\pi) \longrightarrow M - \mathrm{Sing}B_\pi$$

is a branched covering of non-singular type.

PROOF. Note that $\pi : X - \pi^{-1}(\mathrm{Sing}B_\pi) \longrightarrow M - \mathrm{Sing}B_\pi$ is a branched covering whose branch locus is $B_\pi - \mathrm{Sing}B_\pi$. Hence this covering is of non-singular type by Corollary 1.1.9. □

A branched covering $\pi : X \longrightarrow M$ is not necessarily a proper map. Hence, we cannot apply $\pi : X \longrightarrow M$ the proper mapping theorem.

PROPOSITION 1.1.11 Let $\pi : X \longrightarrow M$ be a Galois covering of M. If $\hat{S}$ is a closed complex subspace of X which is G_π-invariant, i.e. $\sigma\hat{S} = \hat{S}$ for all $\sigma \in G_\pi$, then $S = \pi(\hat{S})$ is a closed complex subspace of M. ($\hat{S} = \pi^{-1}(S)$.)

PROOF. For any point $p \in M$, let W be a connected open neighbourhood of p which satisfies the condition of (4) of Definition 1.1.1. For a point $q \in \pi^{-1}(p)$, let U be the connected component of $\pi^{-1}(W)$ with $q \in U$. Since $\pi_U : U \longrightarrow W$ is proper, $\pi(\hat{S} \cap U)$ is a closed complex subspace of W. Since π is a Galois covering and $\hat{S}$ is G_π-invariant, we have $S \cap W = \pi(\hat{S} \cap U)$. Hence S is a closed complex subspace of M. □

PROPOSITION 1.1.12 Let $\pi : X \longrightarrow M$ be a branched covering and C be an irreducible component of $\pi^{-1}(B_\pi)$. Then $\pi(C)$ is an irreducible component of B_π.

PROOF. For a point $p \in \pi(C)$, let W be a connected open neighbourhood of p in M which satisfies the condition of (4) of Definition 1.1.1. For a point $q \in \pi^{-1}(p) \cap C$, let U be the connected component of $\pi^{-1}(W)$ which contains q. Since $\pi_U = \pi|U : U \longrightarrow W$ is proper, $\pi(C \cap U)$ is a complex sub-space of W and is contained in $B_\pi \cap W$. Since π_U is finite, the image $\pi_U(C_1)$, for every irreducible component C_1 of $C \cap U$, is an irreducible component of $B_\pi \cap W$. Hence $\pi_U(C \cap U)$ is a union of some irreducible components of $B_\pi \cap W$. This holds for every point $q \in \pi^{-1}(p) \cap C$. Hence

$$\pi(C) \cap W = \pi(C \cap \pi^{-1}(W)) = \cup_U \pi(C \cap U)$$

is a sum of some irreducible components of $B_\pi \cap W$.

Thus $\pi(C)$ is a hypersurface of M and is irreducible, for C is irreducible. Since $\pi(C) \subset B_\pi$, $\pi(C)$ is in fact an irreducible component of B_π. □

COROLLARY 1.1.13 For every irreducible component C of $\pi^{-1}(B_\pi)$, (1) $C' = C - \pi^{-1}(\mathrm{Sing}B_\pi)$ is open dense in C and (2) the integer e in Theorem 1.1.8 is constant on C'.

The constant e on C' of the corollary is called the *ramification index of*

π *at* C. If $e = 1$, then we say that π *does not ramify at* C. For any irreducible hypersurface E of X which is *not* contained in $\pi^{-1}(B_\pi)$, we define the ramification index of π at E as 1, for convenience.

A similar proof to that of Theorem 1.1.8 shows

THEOREM 1.1.14 Let $\pi : X \longrightarrow M$ be a branched covering of M. Let p and q be points of B_π and $\pi^{-1}(B_\pi)$, respectively such that (1) B_π is normally crossing at p, (2) $q \in \pi^{-1}(p)$, (3) X is non-singular at q and (4) $\pi^{-1}(B_\pi)$ is normally crossing at q. Let W be a sufficiently small connected open neighbourhood of p with a coordinate system $(w_1,\ldots,w_n)$ such that $p = (0,\ldots,0)$ and

$$B_\pi \cap W = \{(w_1,\ldots,w_n) \in W \mid w_k \ldots w_n = 0\}$$

for some k with $1 \leq k \leq n$. Then there is a coordinate system $(z_1,\ldots,z_n)$ in the connected component U of $\pi^{-1}(W)$ with $q \in U$ such that $q = (0,\ldots,0)$,

$$\pi^{-1}(B_\pi) \cap U = \{(z_1,\ldots,z_n) \in U \mid z_k \ldots z_n = 0\} \quad \text{and}$$

$$\pi_U : (z_1,\ldots,z_n) \in U \longmapsto (w_1,\ldots,w_n) = (z_1,\ldots,z_{k-1}, z_k^{e_k},\ldots,z_n^{e_n}) \in W,$$

where e_j, $(k \leq j \leq n)$, is the ramification index of the irreducible component C_j of $\pi^{-1}(B_\pi)$ such that $C_j \cap U = \{(z_1,\ldots,z_n) \in U \mid z_j = 0\}$.

Now let M be an n-dimensional complex manifold and B be a hypersurface of M. Suppose that B has only finite number of irreducible components $D_1,\ldots,D_s$:

$$B = D_1 \cup \ldots \cup D_s.$$

Let $e_1,\ldots,e_s$ be positive integers. Consider the positive divisor

$$D = e_1 D_1 + \ldots + e_s D_s$$

of M.

DEFINITION 1.1.15 A branched covering $\pi : X \longrightarrow M$ is said to *branch at* D (resp. *branch at most at* D), if (1) $B_\pi = B$ (resp. $B_\pi \subset B$), (2) $R_\pi = \pi^{-1}(B)$ (resp. $R_\pi \subset \pi^{-1}(B)$) and (3) for every irreducible component C of $\pi^{-1}(B)$ with $\pi(C) = D_j$, the ramification index of π at C is e_j (resp. divides e_j).

For branched coverings $\pi : X \longrightarrow M$ and $\pi' : X' \longrightarrow M$ of M, we write $\pi \geq \pi'$ or $\pi' \leq \pi$ if there is a morphism φ of π to π'.

LEMMA 1.1.16 (1) If $\pi \geq \pi'$ and $\pi' \geq \pi''$, then $\pi \geq \pi''$. (2) If π is a Galois (or finite) covering and if $\pi \geq \pi'$ and $\pi \leq \pi'$, then π and π' are equivalent. (3) If π is a covering which branches (at most) at D and if $\pi' \leq \pi$, then π' branches at most at D.

PROOF. (1) is trivial. We prove (2). Let φ and φ' be morphisms of π to π' and π' to π, respectively. Then $\sigma = \varphi'.\varphi$ is a morphism of π to itself. That is, σ is a surjective holomorphic map of X onto itself such that the diagram

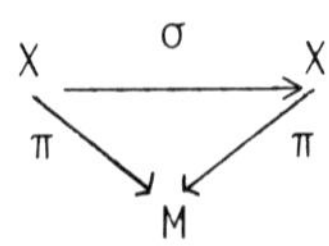

commutes. If π is finite, then σ maps each fibre of π bijectively onto itself. Hence, σ is a covering transformation of π. Hence φ is an isomorphism of π to π'. Next, suppose that π is a Galois covering. Take points $p \in M - B_\pi$ and $q \in \pi^{-1}(p)$. There is a covering transformation $\tau \in G_\pi$ such that $\tau(q) = \sigma(q)$. Note that σ is a local homeomorphism around q. Hence $\sigma = \tau$ locally. Thus $\sigma = \tau$ globally. This implies that φ is an isomorphism of π to π'.

Finally, we prove (3). Let φ be a morphism of π to π'. For a point $p \in B - \mathrm{Sing}B$, take points $q \in \pi^{-1}(p)$ and $q' = \varphi(q)$. Let W be a connected open neighbourhood of p in M with a coordinate system $(w_1,\ldots,w_n)$ which satisfies the condition of Theorem 1.1.8 with respect to both π and π'. Then π and π' are locally (around q and q', respectively) given by

$$\pi \ : (z_1,\ldots,z_n) \longmapsto (w_1,\ldots,w_n) = (z_1,\ldots,z_{n-1},z_n^e),$$

$$\pi' : (x_1,\ldots,x_n) \longmapsto (w_1,\ldots,w_n) = (x_1,\ldots,x_{n-1},x_n^{e'}),$$

for some positive integers e and e'. φ is then locally given by

$$\varphi \ : (z_1,\ldots,z_n) \longmapsto (x_1,\ldots,x_n) = (z_1,\ldots,z_{n-1},h(z_n)),$$

where $h(z_n)^{e'} = z_n^e$. Since $h(z_n)$ is single valued, e' must divide e and $h(z_n) = \zeta z_n^{e/e'}$ for some constant ζ with $\zeta^{e'} = 1$. Thus $\pi' : X' \longrightarrow M$ branches at most at D. □

THEOREM 1.1.17 Let $\pi : X \longrightarrow M$ be a Galois covering which branches at D. Then (1) for every subgroup H of G_π, there is a branched covering

$$\pi_H : X/H \longrightarrow M$$

such that $\pi_H \leq \pi$, (2) the correspondence $H \longrightarrow \pi' = \pi_H$ is a bijection between subgroups H of G and equivalence classes of branched coverings π' of M such that $\pi' \leq \pi$, and (3) H is normal (resp. of finite index) if and only if π_H is a Galois (resp. finite) covering. G_{π_H} is, in this case, isomorphic to G_π/H.

PROOF. (1) Note that H acts on X properly discontinuously, (see Theorem 1.1.7). We identify M with X/G_π. Then it is clear that the natural map

$$\pi_H : X/H \longrightarrow M = X/G_\pi$$

is a branched covering of M. The natural projection $\varphi : X \longrightarrow X/H$ is then a morphism of π to π_H. Hence $\pi_H \leq \pi$.

(2) Every $\pi' : X' \longrightarrow M$ with $\pi' \leq \pi$ induces an unramified covering

$$\mu' : X' - \pi'^{-1}(B) \longrightarrow M - B.$$

It is well known that the correspondence

$$H \longmapsto \mu' = \mu_H$$

is a bijection between subgroups H of G_π and equivalence classes of unramified coverings μ' of M - B such that $\mu' \leq \mu$, where

$$\mu : X - \pi^{-1}(B) \longrightarrow M - B$$

is the unramified Galois covering induced by π. Note that μ_H is the restriction of π_H to $X/H - \pi_H^{-1}(B)$. Hence the correspondence $H \longmapsto \pi' = \pi_H$ is bijective.

(3) As in the proof of (2), $\pi_H : X/H \longrightarrow M = X/G_\pi$ induces an unramified covering

$$\mu_H : X/H - \pi_H^{-1}(B) \longrightarrow M - B.$$

Hence the assertion is reduced to the well-known similar assertion with respect to the relation between H and μ_H. □

EXAMPLE 1.1.18 Put $X = \mathbb{C}^2$, $M = \mathbb{C}^2$ and

$$\pi : (x,y) \in X \longmapsto (v,w) = (x^2+xy+y^2,\ x^2y+xy^2) \in M.$$

Then π is a finite Galois covering with $G_\pi \cong S_3$, the 3-rd symmetric group, $R_\pi = \{(x,y) \mid (x-y)(x+2y)(2x+y) = 0\}$ and $B_\pi = \{(v,w) \mid v^3/27 = w^2/4\}$. π branches at $D = 2B_\pi$.

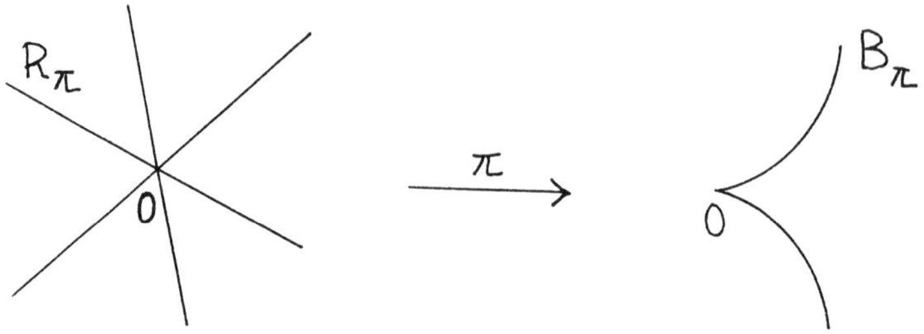

(The origin of B_π is called a *simple cusp of multiplicity* 2. See, for example, Namba [44].)

Consider the subgroups $H = \langle(12)\rangle$ and $N = \langle(123)\rangle$ of G, where the notation $\langle\sigma\rangle$ means the group generated by σ. Then we have the following diagrams which are dual to each other:

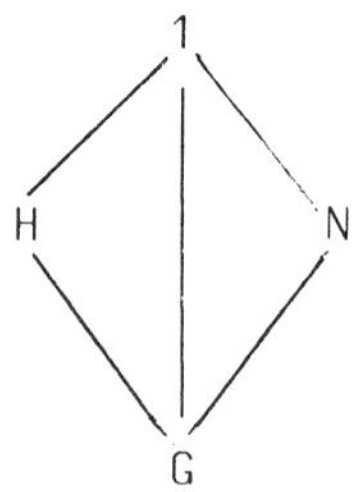

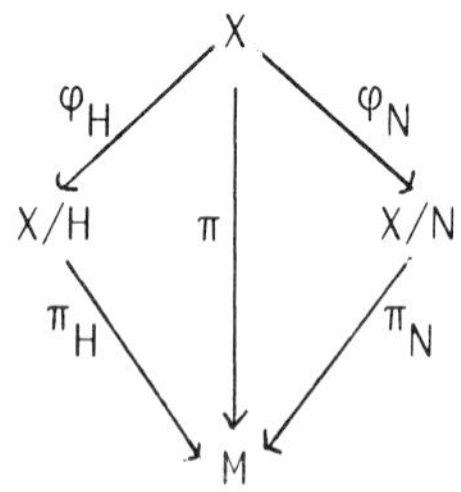

X/H can be then identified with $\mathbb{C}^2$ and φ_H and π_H are given by

$$\varphi_H : (x,y) \in X \longmapsto (s,t) = (x+y,\ xy) \in X/H \quad \text{and}$$

$$\pi_H : (s,t) \in X/H \longmapsto (v,w) = (s^2-t,\ st) \in M.$$

Note that (φ_H is a branched covering and) $R_{\varphi_H} = \{(x,y) \mid x = y\}$ and $B_{\varphi_H} = \{(s,t) \mid t = s^2/4\}$. Note also that

$$R_{\pi_H} = \{(s,t) \mid t+2s^2 = 0\},$$

$$\varphi_H^{-1}(R_{\pi_H}) = \{(x,y) \mid (x+2y)(2x+y) = 0\} \quad \text{and}$$

$$\pi_H^{-1}(B_{\pi_H}) = R_{\pi_H} \cup B_{\varphi_H} .$$

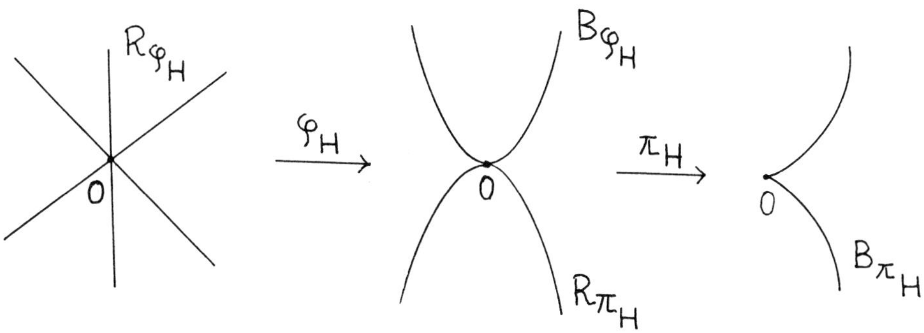

On the other hand, since N is a normal subgroup of G, $\pi_N : X/N \longrightarrow M$ is a Galois covering. X/N can be identified with the surface in $\mathbb{C}^3$ defined by the equation

$$X/N = \{(s,t,u) \in \mathbb{C}^3 \mid u^2-4s^3+27t^2 = 0\}.$$

The origin O = (0,0,0) is the unique singular point of X/N. φ_N and π_N are given by

$$\varphi_N : (x,y) \in X \longmapsto (s,t,u) = (x^2+xy+y^2,\ x^2y+xy^2,\ (x-y)(x+2y)(2x+y)) \in X/N,$$

$$\pi_N : (s,t,u) \in X/N \longmapsto (v,w) = (s,t) \in M.$$

π_N is a 2-fold (cyclic) covering of M. Note that

$$\varphi_N : X - \{0\} \longrightarrow X/N - \{0\}$$

is unramified.

DEFINITION 1.1.19 A Galois covering $\pi : X \longrightarrow M$ which branches at D is said to be *maximal* if, for any covering $\pi' = X' \longrightarrow M$ which branches at most at D, $\pi \geq \pi'$, that is, there is a morphism of π to π'.

By Lemma 1.1.16, a maximal covering is unique up to equivalence, if it exists.

1.2 Construction of maximal branched coverings of non-singular type

Let M be an n-dimensional complex manifold and B be a *non-singular* hypersurface of M which has a finite number of connected components $D_1,\ldots,D_s$:

$$B = D_1 \cup \ldots \cup D_s.$$

Let $e_1,\ldots,e_s$ be positive integers and consider the positive divisor

$$D = e_1D_1 + \ldots + e_sD_s$$

on M.

Following Kato [37], we construct a maximal covering of M which branches at D, under some condition on the pair (M,D). It should be noted, by Corollary 1.1.9, that any covering of M which branches at most at D is necessarily of non-singular type.

We choose a point $p_o \in M-B$, fix it once for all and call it the *base point*. We also choose a point $p_j \in D_j$ for each j, $1 \leq j \leq s$. Let W be an open neighbourhood of p_j in M with a coordinate system $(w_1,\ldots,w_n)$ such that $p_j = (0,\ldots,0)$ and

$$D_j \cap W = B \cap W = \{(w_1,\ldots,w_n) \in W \mid w_n = 0\}.$$

Choose a point $p_j' = (0,\ldots,0,a_j)$, $(a_j \neq 0)$, in $(M-D_j) \cap W$. Consider the circle

$$\alpha_j = \{(0,\ldots,0,a_je^{\sqrt{-1}\vartheta}) \mid 0 \leq \vartheta \leq 2\pi\},$$

oriented in a counterclockwise sense. Let β_j be a path in M-B from p_o to p_j' and put

$$\gamma_j = \beta_j\alpha_j\beta_j^{-1} .$$

(Here, the composition of paths α and β is not written as $\beta\alpha$, but $\alpha\beta$.)

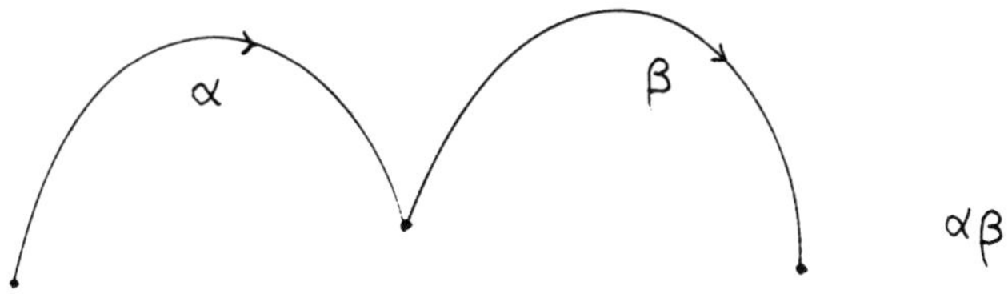

Then γ_j is a closed path in M-B which starts from p_0.

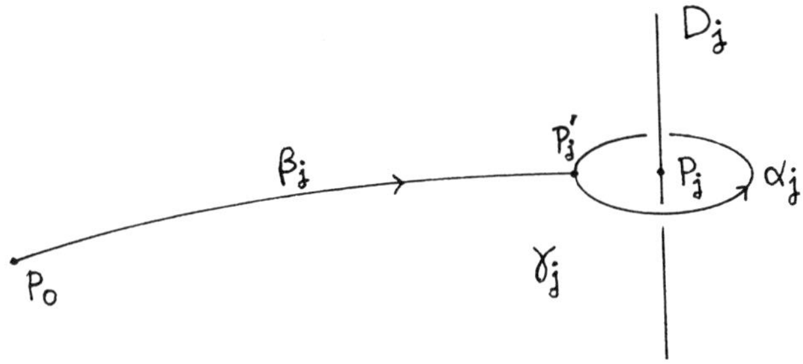

In general, a closed path γ in M-B starting from p_0 determines a homotopy class in the fundamental group $\pi_1(M-B, p_0)$ of M-B which is denoted by γ by abuse of notation.

Let J be the smallest normal subgroup of $\pi_1(M-B, p_0)$ which contains $\gamma_1^{e_1},\dots,\gamma_s^{e_s}$. Put

$$G = \pi_1(M-B, p_0)/J.$$

The following lemma is clear from the definition of J.

<u>LEMMA 1.2.1</u> (1) J does not depend on the choice of the points p_j nor the closed paths γ_j. (2) J is the subgroup of $\pi_1(M-B, p_0)$ generated by $\{\sigma\gamma_j^{e_j}\sigma^{-1} \mid \sigma \in \pi_1(M-B, p_0),\ 1 \le j \le s\}$.

<u>LEMMA 1.2.2</u> Let $\pi : X \longrightarrow M$ be a covering which branches at most at D. Put $K = \pi_*(\pi_1(X-\pi^{-1}(B), q_0))$, where $q_0 \in \pi^{-1}(p_0)$ and $\pi_* : \pi_1(X-\pi^{-1}(B), q_0) \longrightarrow \pi_1(M-B, p_0)$ is the injective homomorphism induced by π. Then $J \subset K$.

<u>PROOF</u>. For any point $q_0' \in \pi^{-1}(p_0)$, let $\hat{\gamma}$ be the path in $X-\pi^{-1}(B)$ with the initial point q_0' such that $\pi(\hat{\gamma}) = \gamma_j^{e_j}$.

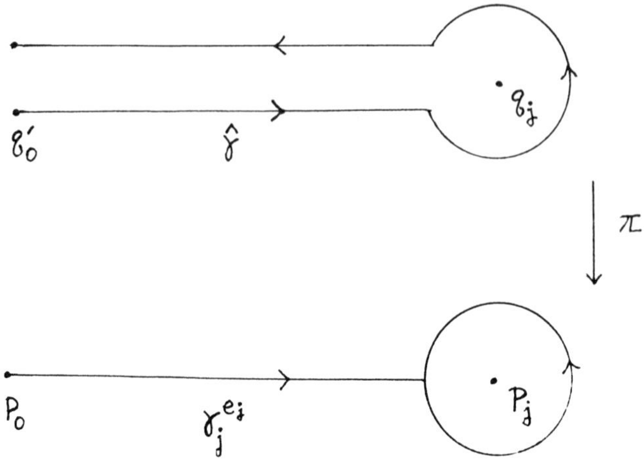

For any point $q_j \in \pi^{-1}(p_j)$, π is locally (around q_j) given by

$$\pi : (z_1, \ldots, z_n) \longmapsto (w_1, \ldots, w_n) = (z_1, \ldots, z_{n-1}, z_n^d),$$

where d divides e_j, (see Theorem 1.1.8). Hence $\hat{\gamma}$ is in fact a *closed* path in $X - \pi^{-1}(B)$ such that $\pi(\hat{\gamma}) = \gamma_j^{e_j}$.

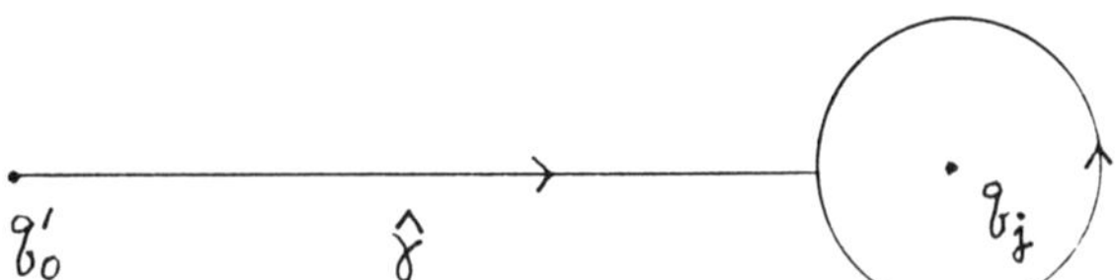

Next, let σ be a closed path in M-B which starts from p_o and $\hat{\sigma}$ be the path in $X - \pi^{-1}(B)$ from q_o to q_o' such that $\pi(\hat{\sigma}) = \sigma$.

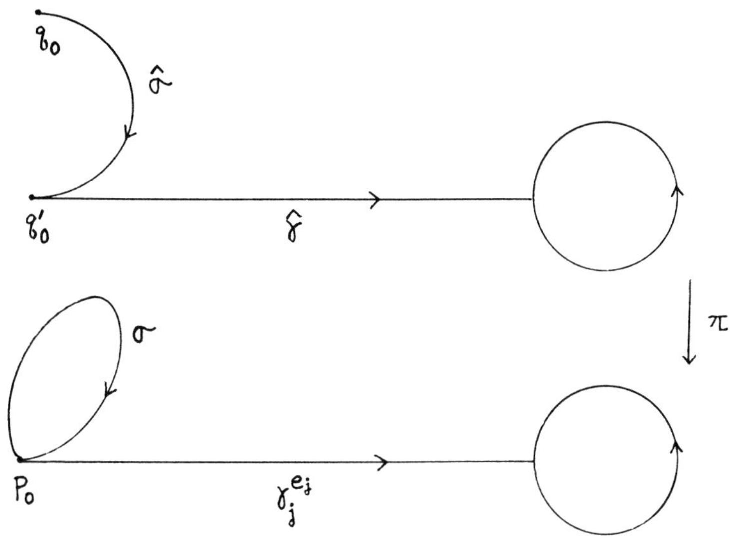

Then $\hat{\sigma}\hat{\gamma}\hat{\sigma}^{-1}$ is a *closed* path in $X - \pi^{-1}(B)$ which starts from q_o such that $\pi(\hat{\sigma}\hat{\gamma}\hat{\sigma}^{-1}) = \sigma\gamma_j^{e_j}\sigma^{-1}$. This shows that $J \subset K$. □

<u>LEMMA 1.2.3</u> Let d_j be the smallest positive integer such that $\gamma_j^{d_j} \in J$. Then (1) d_j divides e_j and (2) the smallest normal subroup of $\pi_1(M-B, p_o)$ which contains $\gamma_1^{d_1}, \ldots, \gamma_s^{d_s}$ is J.

<u>PROOF</u>. (1) Write $e_j = a_j d_j + b_j$, where a_j and b_j are integers such that $0 \le b_j \le d_j - 1$. Then $\gamma_j^{b_j} = \gamma_j^{e_j}(\gamma_j^{d_j})^{-a_j} \in J$. Hence $b_j = 0$.

(2) Let J' be the smallest normal subgroup of $\pi_1(M-B, p_o)$ which contains $\gamma_1^{d_1}, \ldots, \gamma_s^{d_s}$. Since $\gamma_j^{d_j} \in J$ for $1 \le j \le s$, we have $J' \subset J$. On the other hand, since $\gamma_j^{e_j} = (\gamma_j^{d_j})^{a_j} \in J'$, for $1 \le j \le s$, we have $J \subset J'$. Hence $J = J'$. □

Now, we consider the following topological condition on the pair (M,D) :

<u>CONDITION 1.2.4</u> If $\gamma_j^d \in J$, then $d \equiv 0 \pmod{e_j}$, for every j, $1 \le j \le s$.

It is easy to check that this condition is independent of the choice of p_j, p'_j and γ_j.

PROPOSITION 1.2.5 If there is a covering $\pi : X \longrightarrow M$ which branches at D, then Condition 1.2.4 is satisfied for (M,D).

PROOF. By Lemma 1.2.2, we have

$$J \subset K = \pi_*(\pi_1(X-\pi^{-1}(B), q_o)).$$

Hence, if $\pi_j^d \in J$, then $\gamma_j^d \in K$. This means that there is a closed path $\hat{\gamma}$ in $X - \pi^{-1}(B)$ which starts from q_o such that $\pi(\hat{\gamma}) = \gamma_j^d$. For any point $q_j \in \pi^{-1}(p_j)$, π is locally (around q_j) given by

$$\pi : (z_1,\ldots,z_n) \longmapsto (w_1,\ldots,w_n) = (z_1,\ldots,z_{n-1},z_n^{e_j}),$$

(see Theorem 1.1.8). Hence, if $d \not\equiv 0 \pmod{e_j}$, then $\hat{\gamma}$ can not be closed, a contradiction. Thus $d \equiv 0 \pmod{e_j}$. □

Conversely, under Condition 1.2.4, we can construct a maximal covering

$$\tilde{\pi} : \tilde{M} = \tilde{M}(D) \longrightarrow M$$

of non-singular type which branches at D as follows.

Let

$$\mu_o : (M-B)_o \longrightarrow M-B$$

be the universal covering of M-B. Let

$$\mu_1 : ((M-B)_1, q_o) \longrightarrow (M-B, p_o)$$

be the unramified Galois covering of M-B which corresponds to J:

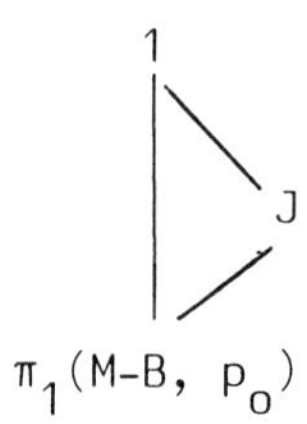

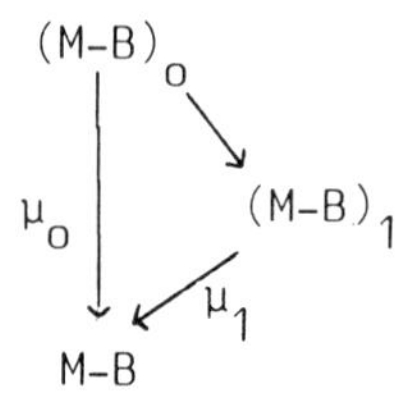

Here $q_o \in \mu_1^{-1}(p_o)$ and the natural injective homomorphism

$$\mu_{1*} : \pi_1((M-B)_1, q_o) \longrightarrow \pi_1(M-B, p_o)$$

has the image J.

LEMMA 1.2.6 Let γ be a closed path in M-B which starts from p_o such that $\gamma \in J$. Then every path $\tilde{\delta}$ in $(M-B)_1$ such that $\mu_1(\tilde{\delta}) = \gamma$, is closed.

PROOF. There is a closed path $\tilde{\gamma}$ in $(M-B)_1$ which starts from q_o such that $\mu_1(\tilde{\gamma}) = \gamma$. Since $\mu_1 : (M-B)_1 \longrightarrow M-B$ is a Galois covering, every path $\tilde{\delta}$ in $(M-B)_1$ such that $\mu_1(\tilde{\delta}) = \gamma$ is also closed. □

Now, take a point $p \in B = D_1 \cup \ldots \cup D_s$. Suppose that p is a point of D_1, say. Take an open neighbourhood W of p with a coordinate system $(w_1,\ldots,w_n)$ such that $p = (0,\ldots,0)$ and $D_1 \cap W = B \cap W = \{w_n = 0\}$. We may identify W and $W - B = W - D_1$ with

$$\mathbb{D} \times \ldots \times \mathbb{D} \quad \text{(n-times) and}$$

$$\mathbb{D} \times \ldots \times \mathbb{D} \times \mathbb{D}^*,$$

respectively, where $\mathbb{D} = \{w \in \mathbb{C} \mid |w| < 1\}$ and $\mathbb{D}^* = \mathbb{D} - \{0\}$. Let U^* be a connected component of $\mu_1^{-1}(W-B)$. Then $\mu_1 : U^* \longrightarrow W-B$ is an unramified covering map. It induces an injective homomorphism

$$\mu_{1*} : \pi_1(U^*,r^*) \longrightarrow \pi_1(W-B,\ r),$$

where $r^* \in U^*$ and $r = \mu_1(r^*) \in W-B$. Note that

$$\pi_1(W-B,\ r) = \pi_1(\mathbb{D} \times \ldots \times \mathbb{D} \times \mathbb{D}^*,\ r) \simeq \alpha^{\mathbb{Z}},$$

where $\alpha^{\mathbb{Z}}$ is the free abelian group generated by the circle

$$\alpha = \{(r_1,\ldots,r_{n-1},r_n e^{i\vartheta}) \mid 0 \leq \vartheta \leq 2\pi\}$$

oriented in a counterclockwise sense, $(r = (r_1,\ldots,r_n))$. If $\tilde{\beta}$ is a path in $(M-B)_1$ from q_o to r^*, then $\beta = \mu_1(\tilde{\beta})$ is a path in M-B from p_o to r and $\beta\alpha\beta^{-1}$ is conjugate to γ_1 in $\pi_1(M-B,\ p_o)$. By Lemma 1.2.6, every path $\tilde{\delta}$ in $(M-B)_1$ such that $\mu_1(\tilde{\delta}) = \beta\alpha^{e_1}\beta^{-1}$ is closed. Hence every path $\tilde{\varepsilon}$ in U^* such that $\mu_1(\tilde{\varepsilon}) = \alpha^{e_1}$ is closed. Hence $\mu_1 : U^* \longrightarrow W-B$ is a finite covering and $\pi_1(U^*,r^*)$ is isomorphic under μ_{1*} to $(\alpha^d)^{\mathbb{Z}}$ for some positive integer d which divides e_1;

$$(\alpha^{e_1})^{\mathbb{Z}} \subset \mu_{1*}(\pi_1(U^*,r^*)) = (\alpha^d)^{\mathbb{Z}}.$$

Let $\tilde{\alpha}$ be the closed path in U^* starting from r^* such that $\mu_1(\tilde{\alpha}) = \alpha^d$. Then $\tilde{\beta}\tilde{\alpha}\tilde{\beta}^{-1}$ is a closed path in $(M-B)_1$ which starts from q_o. Hence

$$\mu_{1*}(\tilde{\beta}\tilde{\alpha}\tilde{\beta}^{-1}) = \beta\alpha^d\beta^{-1} = (\beta\alpha\beta^{-1})^d \in J.$$

Since $\beta\alpha\beta^{-1}$ is conjugate to γ_1 in $\pi_1(M-B,\ p_o)$, we have, by Condition 1.2.4,

$$d = e_1 \quad \text{and}$$

$$\mu_{1*}(\pi_1(U^*,r^*)) = (\alpha^{e_1})^{\mathbb{Z}}.$$

Thus there is a biholomorphic map

$$\varphi : U^* \longrightarrow \mathbf{D} \times \dots \times \mathbf{D} \times \mathbf{D}^*$$

such that the diagram

$$\begin{array}{ccc} U^* & \xrightarrow{\varphi} & \mathbf{D} \times \dots \times \mathbf{D} \times \mathbf{D}^* \\ {\scriptstyle \mu_1}\downarrow & & \downarrow{\scriptstyle \nu_1} \\ W-B & = & \mathbf{D} \times \dots \times \mathbf{D} \times \mathbf{D}^* \end{array}$$

commutes, where ν_1 is given by

$$\nu_1 : (z_1,\dots,z_n) \longmapsto (w_1,\dots,w_n) = (z_1,\dots,z_{n-1},z_n^{e_1}).$$

We identify U^* (resp. μ_1) with $\mathbf{D} \times \dots \times \mathbf{D} \times \mathbf{D}^*$ (resp. ν_1) through φ. Put

$$U = \mathbf{D} \times \dots \times \mathbf{D} \quad (\supset U^*)$$

and extend μ_1 to $\tilde{\pi} : U \longrightarrow W$ as follows.

$$\tilde{\pi} : (z_1,\dots,z_n) \in U \longmapsto (w_1,\dots,w_n) = (z_1,\dots,z_{n-1},z_n^{e_1}) \in W.$$

Consider the union

$$(M-B)_1 \cup (\cup U)$$

where $\cup U$ runs over all $p \in B$ and all connected components U^* of $\mu_1^{-1}(W-B)$. Then, through the natural inclusions $U^* \subset (M-B)_1$, $(M-B)_1$ and U's can be naturally patched up and define a complex manifold $\tilde{M} = \tilde{M}(D)$. (We can easily check that $\tilde{M}$ is a Hausdorff space.) Also

$$\mu_1 : (M-B)_1 \longrightarrow M-B \quad \text{and}$$

$$\tilde{\pi} : U \longrightarrow W$$

can be patched up and define a global surjective holomorphic map

$$\tilde{\pi} : \tilde{M} - \tilde{M}(D) \longrightarrow M.$$

By the construction, it is clear that $\tilde{\pi} : \tilde{M} \longrightarrow M$ is a covering which branches at D. By Theorem 1.1.7, this is a Galois covering with the covering transformation group $G = \pi_1(M-B, p_o)/J$.

If $\pi : X \longrightarrow M$ is a covering which branches at most at D, then $K = \pi_*(\pi_1(X-\pi^{-1}(B), q_o)) \supset J$, by Lemma 1.2.2. Hence there is a morphism φ_1 of $\mu_1 : (M-B)_1 \longrightarrow M-B$ to $X - \pi^{-1}(B) \longrightarrow M-B$. By Proposition 1.1.6, φ_1 can be extended to a morphism φ of $\tilde{\pi} : \tilde{M} \longrightarrow M$ to $\pi : X \longrightarrow M$. Thus $\tilde{\pi} : \tilde{M} \longrightarrow M$ is a maximal covering which branches at D. Note that $\varphi : \tilde{M} \longrightarrow X$ is a Galois covering of non-singular type such that $\pi\varphi = \tilde{\pi}$.

We show that $\tilde{M} = \tilde{M}(D)$ is simply connected. In fact, let $\tilde{\gamma}$ be a closed path in $\tilde{M}$ which starts from q_o. By deforming if necessary, we may assume that $\tilde{\gamma}$ is contained in $(M-B)_1$. Put $\gamma = \mu_1(\tilde{\gamma})$. Then the homotopy class γ can be written as a product of the elements of

$$\{\delta\gamma_j^{e_j}\delta^{-1},\ \delta\gamma_j^{-e_j}\delta^{-1} \mid 1 \leq j \leq s,\ \delta \text{ is a closed path in } M\text{-}B \text{ which starts from } p_o\}.$$

Let $\tilde{\delta}$ be the path in $(M-B)_1$ which starts from q_o such that $\mu_1(\tilde{\delta}) = \delta$. The end point q_1 of $\tilde{\delta}$ is in $\mu_1^{-1}(p_o)$. Let $\tilde{\gamma}_j$ be the path in $(M-B)_1$ which starts from q_1 such that $\mu_1(\tilde{\gamma}_j) = \gamma_j^{e_j}$. Then, by Lemma 1.2.6, $\tilde{\gamma}_j$ is a closed path. Hence $\tilde{\delta}\tilde{\gamma}_j\tilde{\delta}^{-1}$ and $\tilde{\delta}\tilde{\gamma}_j^{-1}\tilde{\delta}^{-1}$ $(1 \leq j \leq s)$ are closed paths. $\tilde{\gamma}$ is a finite product of the homotopy classes of them. By the construction of $\tilde{M}$, $\tilde{\gamma}_j$ is homotopic to zero in $\tilde{M}$. Hence $\tilde{\gamma}$ is also homotopic to zero. We conclude

<u>THEOREM 1.2.7</u> Under Condition 1.2.4, there is a maximal covering $\tilde{\pi} : \tilde{M} = \tilde{M}(D) \longrightarrow M$ which branches at D. $\tilde{M}$ is simply connected. Moreover, there is a natural one-to-one correspondence between (resp. normal, resp. of finite index) subgroups H of $G = \pi_1(M-B, p_o)/J$ and equivalence classes of (resp. Galois, resp. finite) coverings $\pi : X \longrightarrow M$ which branch at most at D. Here, $\pi : X \longrightarrow M$ branches at D if and only if the following condition for K $(H = K/J)$ is satisfied: if $\gamma_j^d \in K$, then $d \equiv 0 \pmod{e_j}$ for $1 \leq j \leq s$.

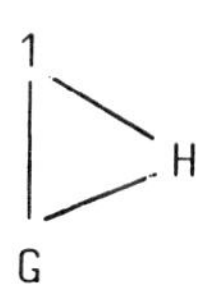

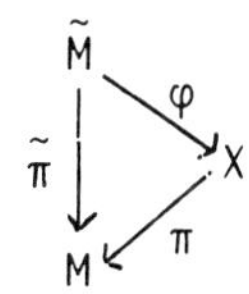

PROOF. The first part of the theorem has been already proved. The one-to-one correspondence $H \longleftrightarrow \pi$ is from Theorem 1.1.7. If $\pi : X \longrightarrow M$ branches at D, then, around $q_j \in \pi^{-1}(p_j)$, π is given by

$$\pi : (z_1,\ldots,z_n) \longmapsto (w_1,\ldots,w_n) = (z_1,\ldots,z_{n-1},z_n^{e_j}).$$

Hence, if $\gamma_j^d \in K$, then $d \equiv 0 \pmod{e_j}$. Conversely, if $\pi : X \longrightarrow M$ branches at most at D and does not branch at D, then we may assume that, around $q_1 \in \pi^{-1}(p_1)$, say, π is locally given by

$$\pi : (z_1,\ldots,z_n) \longmapsto (w_1,\ldots,w_n) = (z_1,\ldots,z_{n-1},z_n^d),$$

where $d \neq e_1$ and d divides e_1. Then, clearly, $\gamma_1^d \in K$, so the condition is not satisfied. □

If we put

$$e_1 = \ldots = e_s = 1 \quad \text{and} \quad D = D_1 + \ldots + D_s,$$

then Condition 1.2.4 is automatically satisfied. In this case, $\tilde{\pi} : \tilde{M} = \tilde{M}(D) \longrightarrow M$ is unramified. Since $\tilde{M}$ is simply connected, $\tilde{\pi} : \tilde{M} \longrightarrow M$ is nothing but the universal covering

$$\pi_o : M_o \longrightarrow M.$$

Its covering transformation group can be identified with $\pi_1(M, p_o)$. Thus we have the following corollary which is intuitively clear. (See Fox [15, p.254].)

COROLLARY 1.2.8 Let J_o be the smallest normal subgroup of $\pi_1(M-B, p_o)$ which contains $\gamma_1,\ldots,\gamma_s$. Then there is a natural exact sequence of groups :

$$1 \longrightarrow J_o \longrightarrow \pi_1(M-B, p_o) \longrightarrow \pi_1(M, p_o) \longrightarrow 1.$$

Under Condition 1.2.4 again, since $\pi_o : M_o \longrightarrow M$ is a covering which branches at most at $D = e_1D_1 + \ldots + e_sD_s$, there is a morphism φ of $\tilde{\pi}$ to π_o. Thus

COROLLARY 1.2.9 Under Condition 1.2.4, there is a Galois covering $\varphi : \tilde{M}(D) \longrightarrow M_o$ of non-singular type such that $\pi_o\varphi = \tilde{\pi}$ and $G_\varphi \simeq J_o/J$.

The simply connectedness of $\tilde{M}$ characterizes $\tilde{\pi} : \tilde{M} \longrightarrow M$. In fact,

THEOREM 1.2.10 Let $\pi : X \longrightarrow M$ be a covering which branches at D such that X is simply connected. Then $\pi : X \longrightarrow M$ is a maximal covering which branches at D.

PROOF. By Proposition 1.2.5, Condition 1.2.4 is satisfied for (M,D). There is a Galois covering $\varphi : \tilde{M} \longrightarrow X$ such that $\pi\varphi = \tilde{\pi}$. Since $\pi : X \longrightarrow M$ branches at D, φ is unramified. Since X is simply connected, φ is biholomorphic. Thus φ is an isomorphism of $\tilde{\pi}$ to π. □

COROLLARY 1.2.11 Let $\pi : X \longrightarrow M$ be a covering which branches at D. Let $\psi_o : X_o \longrightarrow X$ be the universal covering of X. Then $\pi\psi_o : X_o \longrightarrow M$ is a maximal covering which branches at D.

In the rest of this section, let M *be a compact Riemann surface of genus* g. We apply the above discussion to this case.

It is well known that M can be recovered by glueing the sides together in pairs of the 4g-gon as in the figure:

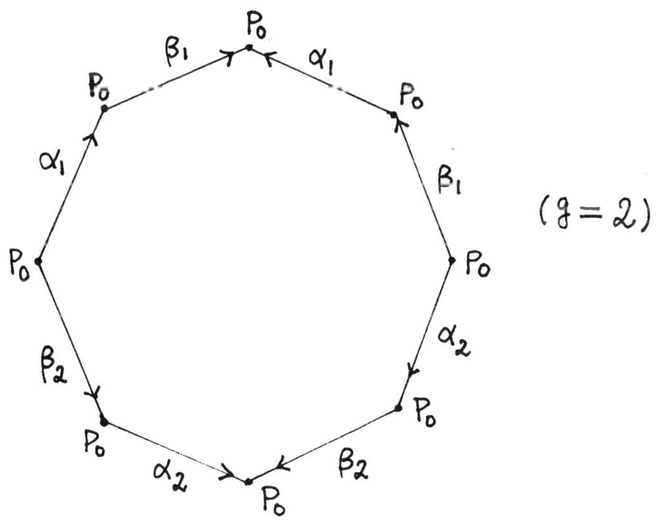

The divisor D_j is in this case a point divisor p_j. Thus

$$B = \{p_1, \ldots, p_s\},$$

$$D = e_1 p_1 + \ldots + e_s p_s.$$

It is well known that the fundamental group $\pi_1(M-B, \ p_o)$ is generated by $\alpha_1, \beta_1, \ldots, \alpha_g, \beta_g, \gamma_1, \ldots, \gamma_s$ with the unique relation:

$$\alpha_1\beta_1\alpha_1^{-1}\beta_1^{-1}\cdots\alpha_g\beta_g\alpha_g^{-1}\beta_g^{-1}\gamma_1\cdots\gamma_s = 1.$$

(See the figure below.) Let J be the smallest normal subgroup of $\pi_1(M-B, p_0)$ which contains $\gamma_1^{e_1},\ldots,\gamma_s^{e_s}$.

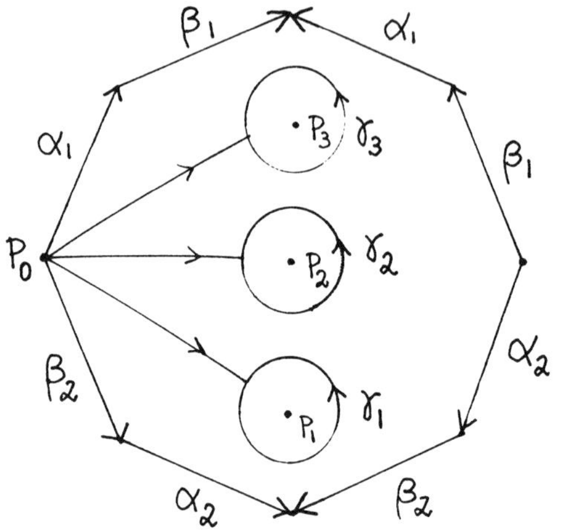

$\begin{pmatrix} g = 2 \\ s = 3 \end{pmatrix}$

$G = \pi_1(M-B, p_0)/J$ is then the group generated by

$$\alpha_1,\beta_1,\ldots,\alpha_g,\beta_g,\gamma_1,\ldots,\gamma_s$$

with the relations

$$\alpha_1\beta_1\alpha_1^{-1}\beta_1^{-1}\cdots\alpha_g\beta_g\alpha_g^{-1}\beta_g^{-1}\gamma_1\cdots\gamma_s = 1 \quad \text{and}$$

$$\gamma_1^{e_1} = 1,\ldots,\gamma_s^{e_s} = 1.$$

Such an (abstract) group G was classically called an F-group. Suppose that $s \geq 1$ and $e_1 \geq 2,\ldots,e_s \geq 2$. Then it can be shown that Condition 1.2.4 is satisfied, *except* for the cases (i) $g = 0$ and $s = 1$ and (ii) $g = 0$, $s = 2$ and $e_1 \neq e_2$. In fact, a stronger assertion holds:

PROPOSITION 1.2.12 Let $s \geq 1$ and $e_1 \geq 2,\ldots,e_s \geq 2$. Suppose that either (1) $g \geq 1$ or (2) $g = 0$, $s = 2$ and $e_1 = e_2$ or (3) $g = 0$ and $s \geq 3$. Then there is a normal subgroup N of G of finite index with the following condition: if $\gamma_j^d \in N$, then $d \equiv 0 \pmod{e_j}$ for $1 \leq j \leq s$.

PROOF. Case 1. $g \geq 1$ (Bundgaard-Nielsen [4]). In this case, we first show that, for every fixed j ($1 \leq j \leq s$), there is a normal subgroup N_j of G of finite index such that, if $\gamma_j^d \in N_j$, then $d \equiv 0 \pmod{e_j}$. Consider the case $j = 1$, say. Let

$$Z = \langle a_1, b_1, \ldots, a_g, b_g, c_1, \ldots, c_s \rangle$$

be the free group generated by $a_1, b_1, \ldots, a_g, b_g, c_1, \ldots, c_s$. Let Z_o be the smallest normal subgroup of Z which contains

$$a_1 b_1 a_1^{-1} b_1^{-1} \ldots a_g b_g a_g^{-1} b_g^{-1} c_1 \ldots c_s, \quad c_1^{e_1}, \ldots, c_s^{e_s}.$$

Let

$$\varphi : Z \longrightarrow G$$

be the surjective homomorphism defined by

$$\varphi(a_k) = \alpha_k, \ \varphi(b_k) = \beta_k \ (1 \leq k \leq g), \ \varphi(c_j) = \gamma_j \quad (1 \leq j \leq s).$$

Then $\ker \varphi = Z_o$ and $Z/Z_o \cong G$. Consider the dihedral group

$$D_{2e_1} = \langle a, b \mid a^2 = b^2 = (ab)^{2e_1} = 1 \rangle.$$

(Here, the notation $\langle a, b \mid \ldots \rangle$ means a group generated by a and b with the relations '...'.) This is a non-abelian finite group of order $4e_1$. Put

$$d_1 = a_1 b_1 a_1^{-1} b_1^{-1} c_1.$$

Then, since $c_1 = b_1 a_1 b_1^{-1} a_1^{-1} d_1$, Z can be regarded as the free group generated by $a_1, b_1, \ldots, a_g, b_g, d_1, c_2, \ldots, c_s$. Consider the surjective homomorphism

$$\psi : Z \longrightarrow D_{2e_1}$$

defined by

$$\psi(a_1) = a, \ \psi(b_1) = b, \ \psi(a_k) = \psi(b_k) = 1 \quad (2 \leq k \leq g),$$

$$\psi(d_1) = 1, \ \psi(c_j) = 1 \quad (2 \leq j \leq s).$$

Then $Z_1 = \ker \psi$ contains $Z_o = \ker \varphi$. In fact

$$\psi(a_1 b_1 a_1^{-1} b_1^{-1} \ldots a_g b_g a_g^{-1} b_g^{-1} c_1 \ldots c_s) = \psi(a_1 b_1 a_1^{-1} b_1^{-1})\psi(c_1)$$
$$= \psi(a_1 b_1 a_1^{-1} b_1^{-1} c_1) = \psi(d_1) = 1, \quad \text{and}$$

$$\psi(c_1^{e_1}) = \psi((a_1 b_1 a_1^{-1} b_1^{-1})^{-1} d_1)^{e_1} = \psi(a_1 b_1 a_1^{-1} b_1^{-1})^{-e_1}$$
$$= (aba^{-1}b^{-1})^{-e_1} = (abab)^{-e_1} = (ab)^{-2e_1} = 1, \quad \text{and}$$

$$\psi(c_j^{e_j}) = 1 \quad \text{for} \quad 2 \leq j \leq s.$$

Hence $Z_1 \supset Z_o$. Put

$$N_1 = Z_1/Z_o.$$

Then N_1 is a normal subgroup of $Z/Z_o \simeq G$. N_1 is of finite index in Z/Z_o. In fact,

$$(Z/Z_o)/N_1 \simeq Z/Z_1 \approx D_{2e_1}.$$

Moreover, if $c_1^d \in Z_1$, then

$$1 = \psi(c_1)^d = (aba^{-1}b^{-1})^{-d} = (ab)^{-2d}.$$

Hence $d \equiv 0 \pmod{e_1}$.

Thus, if we regard N_1 as a normal subgroup of G, then N_1 satisfies the above condition. In a similar way, we can find normal subgroups $N_2,\ldots,N_s$ of G of finite index with the above conditions. Now

$$N = N_1 \cap \ldots \cap N_s$$

is a normal subgroup of G which satisfies the requirement.

<u>Case 2. $g = 0$ (Fox [14]).</u> In this case,

$$G = \langle \gamma_1,\ldots,\gamma_s \mid \gamma_1\cdots\gamma_s = 1,\ \gamma_1^{e_1} = \ldots = \gamma_s^{e_s} = 1\rangle.$$

If $s = 2$ and $e_1 = e_2$, then the assertion is trivial.

Suppose that $s \geq 3$. Then the assertion can be reduced to the following lemma, for the proof of which, we refer the reader to Fox [14].

<u>LEMMA 1.2.13</u> For any integers a, b and c greater than 1, there are permutations A and B of order a and b respectively such that AB has the order c.

In fact, by this lemma, there are, for each j, $2 \leq j \leq s-1$, permutations A_j and B_j of order e_{j-1} and e_j, respectively, such that A_jB_j has the order e_{j+1}. Let

$$\varphi_j : G \longrightarrow \text{a permutation group}$$

be the homomorphism defined by

$$\varphi_j(\gamma_{j-1}) = A_j,\ \varphi_j(\gamma_j) = B_j,\ \varphi_j(\gamma_{j+1}) = (A_jB_j)^{-1},$$
$$\varphi_j(\gamma_k) = 1 \quad \text{for} \quad k \neq j-1,\quad k \neq j,\quad k \neq j+1.$$

Put $N_j = \ker \varphi_j$ and

$$N = N_2 \cap \ldots \cap N_{s-1}.$$

Then N is a normal subgroup of G of finite index.

Moreover, if $\gamma_1^d \in N$, then $\gamma_1^d \in N_2$. Hence

$$1 = \varphi_2(\gamma_1^d) = A_2^d.$$

Thus $d \equiv 0 \pmod{e_1}$. In a similar way, if $\gamma_j^d \in N$, then $d \equiv 0 \pmod{e_j}$ for $2 \leq j \leq s$. Thus N satisfies the requirement. □

This proposition clearly implies Condition 1.2.4 for $D = e_1p_1+\ldots+e_sp_s$, except the cases (i) $g = 0$ and $s = 1$ and (ii) $g = 0$, $s = 2$ and $e_2 \neq e_2$. It is clear that Condition 1.2.4 is not satisfied in these exceptional cases. Hence

THEOREM 1.2.14 The maximal covering $\tilde{\pi} : \tilde{M} = \tilde{M}(D) \longrightarrow M$ *cannot* exist for $D = e_1p_1+\ldots+e_sp_s$ if and only if either (i) $g = 0$ and $s = 1$ or (ii) $g = 0$, $s = 2$ and $e_1 \neq e_2$.

Moreover, Proposition 1.2.12 and Theorem 1.2.7 imply

THEOREM 1.2.15 (Bundgaard-Nielsen-Fox). There is *no finite* Galois covering $\pi : X \longrightarrow M$ which branches at $D = e_1p_1+\ldots+e_sp_s$ if and only if either (i) $g = 0$ and $s = 1$ or (ii) $g = 0$, $s = 2$ and $e_1 \neq e_2$.

Let $\pi : X \longrightarrow M$ be a finite Galois covering of degree d which branches at $D = e_1p_1+\ldots+e_sp_s$. Then X is a compact Riemann surface and $M \cong X/G_\pi$. The genus $g(X)$ of X can be computed by the *Riemann-Hurwitz formula* (see, for example, Namba [44, Theorem 4.1.4]):

$$2g(X) - 2 = d(2g-2) + d \sum_{j=1}^{s} (1-1/e_j).$$

Let $\psi_o : X_o \longrightarrow X$ be the universal covering of X. Then, by Corollary 1.2.11, $\pi\psi_o : X_o \longrightarrow M$ is a maximal covering which branches at D. Hence

PROPOSITION 1.2.16 M(D) is biholomorphic to $\mathbb{D}$ (the unit disc) or $\mathbb{C}$ or $\mathbb{P}^1$ if and only if $\rho = 2g-2 + \sum_{j=1}^{s} (1-1/e_j) > 0$ or $\rho = 0$, or $\rho < 0$, respectively.

For example, put

$$M = \mathbb{P}^1$$

$$D = 2(\infty) + 2(a_1) + \ldots + 2(a_{2m+1}), \quad (a_j \neq a_k \text{ for } j \neq k),$$

$$X' = \{(1:x:y) \in \mathbb{P}^2 \mid y^2 = (x-a_1) \ldots (x-a_{2m+1})\},$$

$$\pi' : (1:x:y) \in X' \longmapsto x \in \mathbb{P}^1,$$

$$\alpha : X \longrightarrow X' \quad \text{a non-singular model of } X',$$

$$\pi = \pi'\alpha : X \longrightarrow \mathbb{P}^1,$$

where (p) is the point divisor on $\mathbb{P}^1$ determined by $p \in \mathbb{P}^1$. Then π is a double covering of $\mathbb{P}^1$ which branches at D. X has the genus $g(X) = m$. Hence, if $m = 1$, then $\tilde{M}(D) \simeq \mathbb{C}$. If $m \geq 2$, then $\tilde{M}(D) \simeq \mathbb{D}$.

Finally, we consider the case $M = \mathbb{P}^1$ and $g(X) = 0$. In this case, $X \simeq \mathbb{P}^1$, so X is simply connected. Hence $\tilde{M}(D) = X$. Moreover,

$$-2 = d\left\{-2 + \sum_{j=1}^{s} (1-1/e_j)\right\}.$$

Since $-2 + \sum_{j=1}^{s} (1-1/e_j) \geq -2+s/2$, we have $-2 \geq d(-2+s/2)$. Hence we have

$$s \leq 3.$$

We can prove directly that the cases (i) and (ii) in Theorem 1.2.14 cannot occur in this case. In fact, if $s = 1$, then

$$2 = d(1+1/e_1).$$

Hence d is divisible by e_1 : $d = d'e_1$. Hence

$$2 = d + d' = d'(e_1+1) \geq 3d',$$

a contradiction. Thus $s \neq 1$.

If $s = 2$, then, assuming $e_1 \geq e_2$, we have

$$2 = d(1/e_1 + 1/e_2) \geq 2d/e_1.$$

Hence $d = e_1$, and so $d = e_2$. Thus, in this case, we have

$$s = 2, \quad e_1 = e_2 = d = m \quad (\geq 2). \tag{1}$$

Suppose that $s = 3$. Assuming $e_1 \geq e_2 \geq e_3$, we have

$$-2 = d(1 - 1/e_1 - 1/e_2 - 1/e_3) \geq d(1 - 3/e_3).$$

Hence $e_3 = 2$. Hence, again,

$$2 = d(1/e_1 + 1/e_2 - 1/2) \leq d(2/e_2 - 1/2).$$

Hence $e_2 \leq 3$.

If $e_2 = 2$, then $d = 2e_1$. Thus, in this case, we have

$$s = 3, \quad e_1 = m \ (\geq 2), \quad e_2 = 2, \quad e_3 = 2, \quad d = 2m. \tag{2}$$

If $e_2 = 3$, then

$$2 = d(1/e_1 - 1/6).$$

Hence $e_1 \leq 5$. Thus, we have the following three cases:

$$s = 3, \quad e_1 = 3, \quad e_2 = 3, \quad e_3 = 2, \quad d = 12. \tag{3}$$

$$s = 3, \quad e_1 = 4, \quad e_2 = 3, \quad e_3 = 2, \quad d = 24. \tag{4}$$

$$s = 3, \quad e_1 = 5, \quad e_2 = 3, \quad e_3 = 2, \quad d = 60. \tag{5}$$

On the other hand, $G = G_\pi$ is a finite subgroup of $\mathrm{Aut}(\mathbb{P}^1)$. If we regard $\mathbb{P}^1$ as the unit sphere S^2, then G is (conjugate to) a finite subgroup of the group SO(3) of rotations (= the maximal compact subgroup of $\mathrm{Aut}(\mathbb{P}^1)$). It is then well known that G is (conjugate to) one of the following groups:

(1) the cyclic group C_m,

(2) the dihedral group D_m,

(3) the tetrahedral group $\simeq A_4$,

(4) the octahedral group $\simeq S_4$,

(5) the icosahedral group $\simeq A_5$.

Here S_d (resp. A_d) is the d-th symmetric (resp. alternative) group.

	s	e_1	e_2	e_3	d	the type of G
(1)	2	m	m	-	m	C_m $(m \geq 2)$
(2)	3	m	2	2	2m	D_m $(m \geq 2)$
(3)	3	3	3	2	12	A_4
(4)	3	4	3	2	24	S_4
(5)	3	5	3	2	60	A_5

It is known that, for suitable choice of affine coordinate systems on $\mathbb{P}^1$, the map $\pi : \mathbb{P}^1 \longrightarrow \mathbb{P}^1/G$ ($\cong \mathbb{P}^1$) can be given by the rational functions

$$w = z^m \qquad \text{for (1),}$$

$$w = \frac{-(z^m-1)^2}{4z^m} \qquad \text{for (2),}$$

$$w = \left[\frac{z^4 + 2\sqrt{3}z^2 - 1}{z^4 - 2\sqrt{3}z^2 - 1}\right]^3 \qquad \text{for (3),}$$

$$w = \frac{(z^8 + 14z^4 + 1)^3}{108z^4(z^4 - 1)^4} \qquad \text{for (4),}$$

$$w = \frac{(z^{20} - 228z^{15} + 494z^{10} + 228z^5 + 1)^3}{-1728z^5(z^{10} + 11z^5 - 1)^5} \qquad \text{for (5).}$$

Here, π branches at

$D = m(\infty) + m(0)$ for (1),

$D = m(\infty) + 2(0) + 2(1)$ for (2),

$D = 3(\infty) + 3(0) + 2(1)$ for (3),

$D = 4(\infty) + 3(0) + 2(1)$ for (4),

$D = 5(\infty) + 3(0) + 2(1)$ for (5).

(See Hochstadt [31].)

Since we know that G is generated by $\gamma_1, \ldots, \gamma_s$ with the relations:

$$\gamma_1 \cdots \gamma_s = \gamma_1^{e_1} = \ldots = \gamma_s^{e_s} = 1,$$

we have the famous

<u>THEOREM 1.2.17</u> <u>(Coxeter-Moser [7].)</u>

(1) $C_m = \langle \gamma \mid \gamma^m = 1 \rangle$, $(m \geq 2)$,

(2) $D_m = \langle \gamma, \delta \mid \gamma^2 = \delta^2 = (\gamma\delta)^m = 1 \rangle$, $(m \geq 2)$,

(3) $A_4 = \langle \gamma, \delta \mid \gamma^2 = \delta^3 = (\gamma\delta)^3 = 1 \rangle$,

(4) $S_4 = \langle \gamma, \delta \mid \gamma^2 = \delta^3 = (\gamma\delta)^4 = 1 \rangle$,

(5) $A_5 = \langle \gamma, \delta \mid \gamma^2 = \delta^3 = (\gamma\delta)^5 = 1 \rangle$.

1.3 Construction of maximal branched coverings (general case)

Let M be an n-dimensional complex manifold and B be a hypersurface of M which has a finite number of irreducible components $D_1, \ldots, D_s$:

$$B = D_1 \cup \ldots \cup D_s.$$

Let $e_1, \ldots, e_s$ be positive integers and consider the positive divisor

$$D = e_1 D_1 + \ldots + e_s D_s$$

on M.

In this section, we construct a maximal covering of M which branches at D, under some topological condition on the pair (M,D).

Let SingB be the singular locus of B. We put

$$M' = M - \mathrm{Sing}B,$$

$$B' = B - \mathrm{Sing}B,$$

$$D'_j = D_j - \mathrm{Sing}B, \quad 1 \le j \le s,$$

$$D' = e_1 D'_1 + \ldots + e_s D'_s.$$

We choose a point $p_o \in M-B$, fixed it once and for all and call it the *base point*. We choose a point $p_j \in D'_j$ for $1 \le j \le s$. Let γ_j be a closed path in M-B which starts from p_o and goes around D_j once in a counterclockwise direction with the centre p_j as in section 1.2. Let J be, as in section 1.2, the smallest normal subgroup of $\pi_1(M-B, p_o)$ which contains $\gamma_1^{e_1}, \ldots, \gamma_s^{e_s}$. Then, as in section 1.2, we have the following two lemmas.

LEMMA 1.3.1 Let $\pi : X \longrightarrow M$ be a branched covering such that $\pi : X-\pi^{-1}(B) \longrightarrow M-B$ is unramified. Then $\pi : X \longrightarrow M$ branches at most at D if and only if $J \subset \pi_*(\pi_1(X-\pi^{-1}(B), q_o))$, where $q_o \in \pi^{-1}(p_o)$ and $\pi_* : \pi_1(X-\pi^{-1}(B), q_o) \longrightarrow \pi_1(M-B, p_o)$ is the injective homomorphism induced by π.

LEMMA 1.3.2 Let $\pi : X \longrightarrow M$ be a covering which branches at most at D. Then π branches at D if and only if $K = \pi_*(\pi_1(X-\pi^{-1}(B), q_o))$ satisfies the following condition: If $\gamma_j^d \in K$, then $d \equiv 0 \pmod{e_j}$, for $1 \le j \le s$.

We now consider the following condition on a subgroup K of $\pi_1(M-B, p_o)$ such that $J \subset K$:

<u>CONDITION 1.3.3</u> Let K be a subgroup of $\pi_1(M-B, p_o)$ such that $J \subset K$. If $\gamma_j^d \in K$, then $d \equiv 0 \pmod{e_j}$ for $1 \leq j \leq s$.

We clearly have

<u>LEMMA 1.3.4</u> Let K_1 and K_2 be subgroups of $\pi_1(M-B, p_o)$ such that $J \subset K_1 \subset K_2$. If K_2 satisfies Condition 1.3.3, then K_1 also satisfies Condition 1.3.3.

Next, let

$$\pi : X \longrightarrow M$$

be a covering of M which branches at most at D. Put

$$K = \pi_*(\pi_1(X-\pi^{-1}(B), q_o)), \quad (q_o \in \pi^{-1}(p_o)).$$

For any point $p \in \mathrm{Sing}B$, let W be an open neighbourhood of p in M such that, for every connected components U of $\pi^{-1}(W)$, $\pi^{-1}(p) \cap U = \{q\}$, one point, and $\pi : U \longrightarrow W$ is surjective, proper (and hence finite). We may assume that W is a small open ball with respect to a local coordinate system with the centre p. Then, for a fixed point $p_o' \in W-B$, $\pi_1(W-B, p_o')$ is isomorphic to

$$\pi_{1,loc,p}(M-B),$$

the *local fundamental group of* $M-B$ *at* p. Let

$$i_* : \pi_1(W-B, p_o') \longrightarrow \pi_1(M-B, p_o')$$

be the homomorphism induced by the inclusion map $i : W-B \hookrightarrow M-B$. Let β be a path from p_o to p_o'. Then we have the isomorphism

$$\gamma \in \pi_1(M-B, p_o') \longmapsto \beta\gamma\beta^{-1} \in \pi_1(M-B, p_o).$$

We identify $\pi_1(M-B, p_o')$ with $\pi_1(M-B, p_o)$ through this isomorphism.

<u>LEMMA 1.3.5</u> There are the corresponding commutative diagrams:

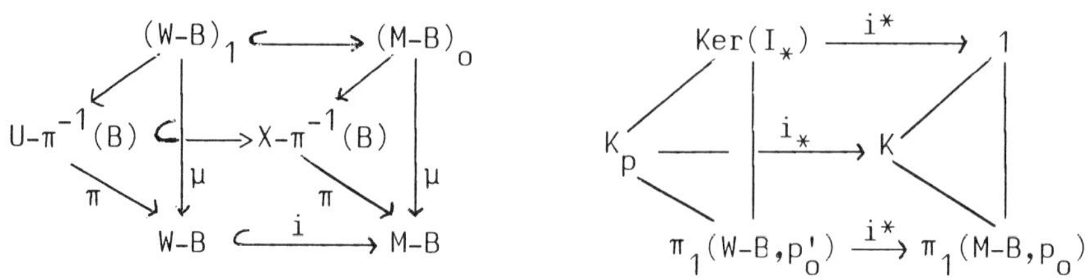

where $\mu: (M-B)_0 \longrightarrow M-B$ is the universal covering, $(W-B)_1$ is a connected component of $\mu^{-1}(W-B)$ and $K_p = i_*^{-1}(K)$.

PROOF Let γ' be a closed path in $U-\pi^{-1}(B)$ which starts from $q_0' \in \pi^{-1}(p_0) \cap U$. Then $i\pi(\gamma')$ can be regarded as an element of K. Hence $\pi(\gamma') \in i_*^{-1}(K) = K_p$. Conversely, if γ is a closed path in $W-B$ which starts from p_0' such that $i_*\gamma \in K$, then the path γ' which starts from q_0' such that $\pi(\gamma') = \gamma$, is closed. Thus

$$\pi_*(\pi_1(U-\pi^{-1}(B), q_0')) = K_p.$$

The other correspondences can be shown in a similar way. □

Note that $\pi : U \longrightarrow W$ is a finite branched covering. Hence K_p is a subgroup of $\pi_1(W-B, p_0')$ of finite index.

Conversely, for a subgroup K of $\pi_1(M-B, p_0)$ with $J \subset K$, we consider the following condition:

CONDITION 1.3.6 Let K be a subgroup of $\pi_1(M-B, p_0)$ such that $J \subset K$. For any point $p \in \mathrm{Sing}B$, $K_p = i_*^{-1}(K)$ is a subgroup of $\pi_1(W-B, p_0')$ of finite index.

Note that $\pi_1(M-B, p_0)$ itself satisfies this condition.

We clearly have

LEMMA 1.3.7 Let K_1 and K_2 be subgroups of $\pi_1(M-B, p_0)$ such that $J \subset K_1 \subset K_2$. If K_1 satisfies Condition 1.3.6, then K_2 also satisfies Condition 1.3.6.

Now we prove

THEOREM 1.3.8 Let K be a (resp. normal, resp. of finite index) subgroup of $\pi_1(M-B, p_0)$ such that $J \subset K$. Suppose that K satisfies Condition 1.3.6. Then there is a unique (up to equivalence) (resp. Galois, resp. finite) covering $\pi : X \longrightarrow M$ which branches at most at D such that

$$K = \pi_*(\pi_1(X-\pi^{-1}(B), q_0)), \quad (q_0 \in \pi^{-1}(p_0)).$$

Moreover, $\pi : X \longrightarrow M$ branches at D if and only if K satisfies Condition 1.3.3.

PROOF. (Following Grothendieck-Raynaud [24, p.340].)

Let $\psi : Y \longrightarrow M-B$ be the unramified covering which corresponds to K, that is,

$$\psi_*(\pi_1(Y, q_o)) = K, \quad (q_o \in \psi^{-1}(p_o)).$$

By the assumption $J \subset K$, a similar proof to that of Theorem 1.2.7 shows that $\psi : Y \longrightarrow M-B$ can be extended to

$$\pi' : X' \longrightarrow M' = M-\mathrm{Sing}B$$

which branches at most at $D' = e_1D_1' + \ldots + e_sD_s'$. (Since K may not be normal in $\pi_1(M-B, p_o)$, we cannot use a similar lemma to Lemma 1.2.6, but we do not need it because, under the notations in the argument after Lemma 1.2.6, $\mu_1 : U^* \longrightarrow W-B$ is a cyclic covering.)

We extend $\pi' : X' \longrightarrow M'$ to a branched covering $\pi : X \longrightarrow M$. For any point $p \in \mathrm{Sing}B$, let W be as before a small open neighbourhood of p in M which is a ball in a coordinate system with the centre p. Let U' be a connected component of $\pi'^{-1}(W')$, where $W' = W-\mathrm{Sing}B$. By Condition 1.3.6, $\pi' : U' \longrightarrow W'$ is a finite branched covering.

By Grauert-Remmert theorem, we can extend π' to a finite branched covering $\pi : U \longrightarrow W$. More precisely, the extension π can be constructed as follows. By the fundamental theorem of Hironaka [29], there is a bimeromorphic holomorphic map

$$f : \hat{W} \longrightarrow W$$

such that (1) $\hat{W}$ is a complex manifold, (2) f maps $f^{-1}(W')$ biholomorphically onto W' and (3) $Z = f^{-1}(B \cap W)$ is a simple normally crossing hypersurface of $\hat{W}$. We identify $f^{-1}(W')$ with W' through f.

For any point $p \in Z$, there is an open neighbourhood V of p in $\hat{W}$ such that $V = \mathbb{D} \times \ldots \times \mathbb{D}$ (n-times) with respect to a coordinate system $(z_1,\ldots,z_n)$ and such that

$$Z \cap V = \{(z_1,\ldots,z_n) \in V \mid z_k\ldots z_n = 0\},$$

$$V - Z = \mathbb{D} \times \ldots \times \mathbb{D} \times \mathbb{D}^* \times \ldots \times \mathbb{D}^*.$$

Since $\pi' : \pi'^{-1}(V-Z) \longrightarrow V-Z$ is a finite unramified covering and since $\pi_1(V-Z, r)$ $(r \in V-Z)$ is isomorphic to $\mathbb{Z} \oplus \ldots \oplus \mathbb{Z}$ (n-k+1 times), there are positive integers $c_k,\ldots,c_n$ and a subgroup H of the covering transformation group T of

$$\begin{array}{ccc} \mathbb{D} \times \ldots \times \mathbb{D} \times \mathbb{D}^* \times \ldots \times \mathbb{D}^* & \xrightarrow{\ \rho\ } & V - Z \\ \cup & & \cup \\ (t_1,\ldots,t_{k-1},t_k,\ldots,t_n) & \longmapsto & (z_1,\ldots,z_n) = (t_1,\ldots,t_{k-1},t_k^{c_k},\ldots,t_n^{c_n}) \end{array}$$

such that $\pi'^{-1}(V-Z) = (\mathbb{D} \times \ldots \times \mathbb{D} \times \mathbb{D}^* \times \ldots \times \mathbb{D}^*)/H$ (identified). T is a finite abelian group which is isomorphic to

$$(\mathbb{Z}/c_k\mathbb{Z}) \oplus \ldots \oplus (\mathbb{Z}/c_n\mathbb{Z}).$$

We have the commutative diagram:

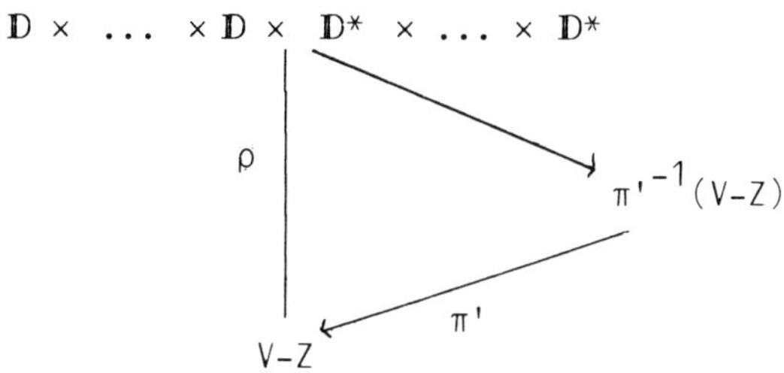

We extend $\mathbb{D} \times \ldots \times \mathbb{D} \times \mathbb{D}^* \times \ldots \times \mathbb{D}^*$ to $\mathbb{D} \times \ldots \times \mathbb{D}$ (n-times) and ρ to

$$\hat{\rho} : \mathbb{D} \times \ldots \times \mathbb{D} \longrightarrow V$$

$$\begin{array}{ccc} \text{Ш} & & \text{Ш} \\ (t_1,\ldots,t_n) & \longmapsto & (z_1,\ldots,z_n) = (t_1,\ldots,t_{k-1},t_k^{c_k},\ldots,t_n^{c_n}) \end{array}$$

We also extend $\pi'^{-1}(V-Z) = (\mathbb{D} \times \ldots \times \mathbb{D} \times \mathbb{D}^* \times \ldots \times \mathbb{D}^*)/H$ to $\hat{V} = (\mathbb{D} \times \ldots \times \mathbb{D})/H$. Then π' can be extended to $\hat{\pi} : \hat{V} \longrightarrow V$ which makes the diagram

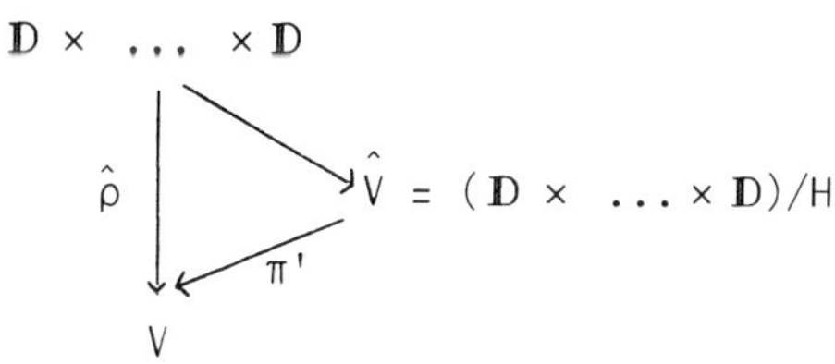

commutative. By Proposition 1.1.5, the extension $\hat{\pi} : \hat{V} \longrightarrow V$ of $\pi' : \pi'^{-1}(V-Z) \longrightarrow V-Z$ is unique (up to equivalence). Hence we can patch them up and get an irreducible normal complex space $\hat{U}$ and a finite branched covering $\hat{\pi} : \hat{U} \longrightarrow \hat{W}$ which extends $\pi' : U' - \pi'^{-1}(B') \longrightarrow W'-B'$. By Proposition 1.1.5 again, the extension is unique (up to equivalence). Hence $\hat{\pi} : \hat{U} \longrightarrow \hat{W}$ is in fact an extension of $\pi' : U' \longrightarrow W'$.

Now, $\hat{\pi}_* \mathcal{O}_{\hat{U}}$ is a finitely generated $\mathcal{O}_{\hat{W}}$-module. Hence $(f.\hat{\pi})_* \mathcal{O}_{\hat{U}}$ is a finitely generated $\mathcal{O}_W$-module. To $(f.\hat{\pi})_*\mathcal{O}_{\hat{U}}$, there corresponds a finite branched covering

$$\pi : U \longrightarrow W,$$

(see, for example, Gunning [26]), which clearly extends $\pi' : U' \longrightarrow W'$.

The extension is unique (up to equivalence). Hence we can patch them and $\pi' : X' \longrightarrow M'$ up and get an irreducible normal complex space X and a covering

$$\pi : X \longrightarrow M$$

which branches at most at D and corresponds to K. By Lemma 1.3.2, π branches at D if and only if K satisfies Condition 1.3.3. □

This theorem asserts that the correspondence

$$\pi \longmapsto K = K(\pi) = \pi_*(\pi_1(X-\pi^{-1}(B), q_o))$$

from the set of all isomorphism classes of coverings $\pi : X \longrightarrow M$ of M which branch at most at D to the set of all subgroups K of $\pi_1(M-B, p_o)$ with $J \subset K$ which satisfy Condition 1.3.6 is one-to-one. The correspondence satisfies:

$$\pi_1 \leq \pi_2 \quad \text{if and only if} \quad K(\pi_1) \supset K(\pi_2).$$

Moreover, this theorem implies the following important theorem:

<u>THEOREM 1.3.9</u> The correspondence $\pi \longmapsto K = K(\pi) = \pi_*(\pi_1(X-\pi^{-1}(B), q_o))$ gives a one-to-one correspondence between isomorphism classes of finite Galois coverings $\pi : X \longrightarrow M$ which branch at most at D and normal subgroups K of $\pi_1(M-B, p_o)$ of finite index such that $J \subset K$. The correspondence satisfies (1) $G_\pi \simeq \pi_1(M-B, p_o)/K(\pi)$, (2) $\pi_1 \leq \pi_2$ if and only if $K(\pi_1) \supset K(\pi_2)$ and (3) π branches at D if and only if $K(\pi)$ satisfies Condition 1.3.3.

<u>PROOF</u>. Let K be a normal subgroup of $\pi_1(M-B, p_o)$ of finite index such that $J \subset K$. Then, under the previous notations,

$$\frac{\pi_1(W-B)}{K_p} \simeq \frac{i_*(\pi_1(W-B))}{K \cap i_*(\pi_1(W-B))} \simeq \frac{K.i_*(\pi_1(W-B))}{K} \subset \frac{\pi_1(M-B,p_o)}{K}.$$

Hence K satisfies Condition 1.3.6. By Theorem 1.3.8, there is a finite Galois covering $\pi : X \longrightarrow M$ which branches at most at D such that $K = \pi_*(\pi_1(X-\pi^{-1}(B), q_o))$. Other assertions of the theorem are trivial. □

Now, put

$$\tilde{K} = \cap K,$$

where $\cap$ runs over all subgroups K of $\pi_1(M-B, p_o)$ with $J \subset K$ which satisfy Condition 1.3.6. Then $\tilde{K}$ is a normal subgroup of $\pi_1(M-B, p_o)$ such that $J \subset \tilde{K}$, but $\tilde{K}$ itself may not satisfy Condition 1.3.6. Put

$$\tilde{G} = \pi_1(M-B, p_o)/\tilde{K}.$$

<u>THEOREM 1.3.10</u> There is a maximal covering $\tilde{\pi} : \tilde{M} = \tilde{M}(D) \longrightarrow M$ which branches at D if and only if $\tilde{K}$ satisfies both Conditions 1.3.3 and 1.3.6. In this case, (1) $\tilde{\pi} : \tilde{M} \longrightarrow M$ corresponds to $\tilde{K}$ and so $G_{\tilde{\pi}} = \tilde{G}$ (identified) and (2) $\tilde{M}$ is simply connected.

<u>PROOF</u>. If $\tilde{\pi} : \tilde{M} \longrightarrow M$ is a maximal covering which branches at D, then $K' = \tilde{\pi}_*(\pi_1(\tilde{M}-\tilde{\pi}^{-1}(B), \tilde{p}_o))$, $(\tilde{p}_o \in \tilde{\pi}^{-1}(p_o))$, satisfies both Conditions 1.3.3 and 1.3.6. By Theorem 1.3.8, for any subgroup K of $\pi_1(M-B, p_o)$ with $J \subset K$ which satisfies Condition 1.3.6, there is a corresponding covering $\pi : X \longrightarrow M$ which branches at most at D. By the maximality, there is a commutative diagram

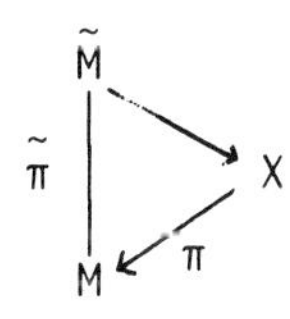

Hence $K' \subset K$. This shows that $K' = \tilde{K}$.

Conversely, suppose that $\tilde{K}$ satisfies both Conditions 1.3.3 and 1.3.6. By Theorem 1.3.8, there is a corresponding covering $\tilde{\pi} : \tilde{M} \longrightarrow M$ which branches at D. By Proposition 1.1.6, this is clearly a maximal covering which branches at D.

Let $\mu : M_o \longrightarrow \tilde{M}$ be the universal covering of $\tilde{M}$. Then $\tilde{\pi}\mu \geq \tilde{\pi}$. On the other hand, $\tilde{\pi}\mu \leq \tilde{\pi}$, because $\tilde{\pi}$ is maximal. Since $\tilde{\pi}$ is a Galois covering, $\tilde{\pi}$ and $\tilde{\pi}\mu$ are equivalent by Lemma 1.1.16. Hence there is a biholomorphic mapping $\mu' : M_o \longrightarrow \tilde{M}$ such that $\tilde{\pi}\mu = \tilde{\pi}\mu'$. This implies that $\mu = \mu'$. Thus $\tilde{M}$ is simply connected. □

In particular

<u>THEOREM 1.3.11</u> Suppose that J satisfies Conditions 1.3.3 and 1.3.6. Then there is a maximal covering $\tilde{\pi} : \tilde{M} = \tilde{M}(D) \longrightarrow M$ which branches at D. In

this case, (1) $\tilde{\pi}$ corresponds to J and so $G_{\tilde{\pi}} = G = \pi_1(M-B, p_0)/J$, (2) $\tilde{\pi}' = \tilde{\pi} \mid \tilde{M}' : \tilde{M}' = \tilde{M} - \tilde{\pi}^{-1}(\text{SingB}) \longrightarrow M' = M-\text{SingB}$ is a maximal covering of non-singular type which branches at D' and (3) $\tilde{M}'$ is simply connected.

PROOF. It suffices to prove (2) and (3). By Condition 1.3.3 and Theorem 1.2.7, there is a maximal covering $\psi : Y \longrightarrow M'$ of M' which branches at D'. Since $\tilde{\pi}' : \tilde{M}' \longrightarrow M'$ corresponds to J, $\tilde{\pi}'$ is equivalent to ψ by Theorem 1.2.7 again. This proves (2). (3) follows from (2) and Theorem 1.2.7. □

COROLLARY 1.3.12 Let $\pi : X \longrightarrow M$ be a covering which branches at D such that $X' = X - \pi^{-1}(\text{SingB})$ is simply connected. Then π is a maximal covering which branches at D such that $G_\pi \simeq \pi_1(M-B, p_0)/J$.

PROOF. Since π branches at D, $K = \pi_*(\pi_1(X-\pi^{-1}(B), q_0))$ satisfies both Conditions 1.3.3 and 1.3.6. By Theorem 1.2.10, $\pi' : X' \longrightarrow M'$ is a maximal covering of M' which branches at D'. Hence K = J by Theorem 1.2.7. □

In this corollary, if X is non-singular and simply connected, then $X - \pi^{-1}(\text{SingB})$ is also simply connected. Hence,

THEOREM 1.3.13 Let X be a complex manifold and $\pi : X \longrightarrow M$ be a covering which branches at D. Then π is maximal if and only if X is simply connected. In this case, $G_\pi \simeq \pi_1(M-B, p_0)/J$.

REMARK 1.3.14 The non-singularity condition of X in the above theorem cannot be dropped. In fact, consider the following example. Put $M = \mathbb{C}^2$ and

$$\begin{array}{ccc} \pi : X = \{(u,v,w) \in \mathbb{C}^3 \mid u^2 = vw\} & \longrightarrow & M \\ \Psi & & \Psi \\ (u,v,w) & \longmapsto & (v,w). \end{array}$$

Then π branches at $D = 2D_1 + 2D_2$ where $D_1 = \{v = 0\}$ and $D_2 = \{w = 0\}$. X is simply connected, because X is a cone. On the other hand, π cannot be maximal. In fact

$$\tilde{\pi} : (x,y) \in \tilde{M} = \mathbb{C}^2 \longmapsto (v,w) = (x^2,y^2) \in M$$

is a maximal covering which branches at D by the above theorem and $\tilde{\pi} = \pi\mu$, where

$$\mu : (x,y) \in \tilde{M} \longmapsto (u,v,w) = (xy,x^2,y^2) \in X.$$

Note that $\mu : \tilde{M} - \{0\} \longrightarrow X - \{0\}$ is unramified and $\tilde{M} - \{0\}$ is simply connected. Hence

$\pi_1(X - \{0\}, q_o) \simeq \mathbb{Z}/2\mathbb{Z}$.

THEOREM 1.3.15 Suppose that B is simple normally crossing. Then there is a maximal covering $\tilde{\pi} : \tilde{M} = \tilde{M}(D) \longrightarrow M$ which branches at D if and only if J satisfies Condition 1.3.3. In this case, $\tilde{K} = J$ and so $G_{\tilde{\pi}} = G = \pi_1(M-B,p_o)/J$.

PROOF. We show that if B is simple normally crossing, then J satisfies Condition 1.3.6. In fact, for any point $p \in \mathrm{Sing}B$, let W be a small open neighbourhood of p with a coordinate system $(w_1,\ldots,w_n)$ such that $p = (0,\ldots,0)$ and

$$B \cap W = \{(w_1,\ldots,w_n) \in W \mid w_k \ldots w_n = 0\}.$$

We may assume that

$$\{(w_1,\ldots,w_n) \in W \mid w_j = 0\} = D_j \cap W \quad \text{for } k \leq j \leq n.$$

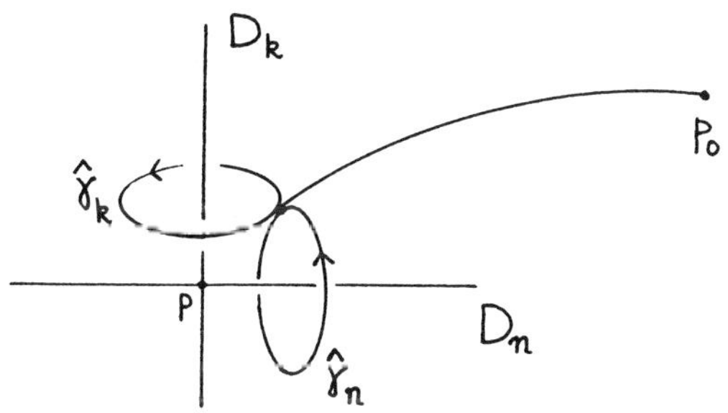

Let $\hat{\gamma}_j$ $(k \leq j \leq n)$ be a closed path in M-B as in the figure, which rounds D_j near p once in a counterclockwise direction. Then (1) $\hat{\gamma}_j$ is conjugate to γ_j and (2) $\hat{\gamma}_j$ $(k \leq j \leq n)$ commute each other. Then, under the notations in Lemma 1.3.5, there is a natural isomorphism

$$\pi_1(W-B) \simeq (\mathbb{Z}\hat{\gamma}_k) \oplus \ldots \oplus (\mathbb{Z}\hat{\gamma}_n).$$

We identify $\pi_1(W-B)$ with $(\mathbb{Z}\hat{\gamma}_k) \oplus \ldots \oplus (\mathbb{Z}\hat{\gamma}_n)$ through this isomorphism. Then we have clearly

$$\mathbb{Z}(e_k\hat{\gamma}_k) + \ldots + \mathbb{Z}(e_n\hat{\gamma}_n) \subset J_p.$$

Hence J_p is a subgroup of $\pi_1(W-B)$ of finite index. Thus J satisfies Condition 1.3.6. This implies that $\tilde{K} = J$. Now the theorem reduces to Theorem 1.3.10. □

COROLLARY 1.3.16 (cf. Kato [37]). Suppose that B is simple normally crossing and J satisfies Condition 1.3.3. Under the notations of the proof of Theorem 1.3.15, suppose that

$$J_p = \mathbb{Z}(e_k\hat{\gamma}_k) + \ldots + \mathbb{Z}(e_n\hat{\gamma}_n)$$

for any point $p \in \mathrm{Sing}B$. Then the maximal covering $\tilde{\pi} : \tilde{M} = \tilde{M}(D) \longrightarrow M$ which branches at D is of simple normally crossing type.

PROOF. As in the proof of Theorem 1.3.8, $\tilde{\pi}^{-1}(W)$ can be identified with $(\mathbb{D} \times \ldots \times \mathbb{D})/H$, where

$$H = J_p/(\mathbb{Z}(e_k\hat{\gamma}_k) + \ldots + \mathbb{Z}(e_n\hat{\gamma}_n)).$$

By the assumption, this is zero. Hence $\tilde{\pi}^{-1}(W) = \mathbb{D} \times \ldots \times \mathbb{D}$ and

$$\begin{array}{cccc} \tilde{\pi} : & \tilde{\pi}^{-1}(W) & \longrightarrow & W \\ & \cup & & \cup \\ & (t_1,\ldots,t_n) & \longmapsto & (w_1,\ldots,w_n) = (t_1,\ldots,t_{k-1},t_k^{e_k},\ldots,t_n^{e_n}). \end{array}$$

Hence $\tilde{M}$ is non-singular and $\tilde{\pi} : \tilde{M} \longrightarrow M$ is of simple normally crossing type. □

The existence theorem on *universal uniformizations of branchfolds* by Kato [37] is much more general than this corollary. The following theorem is also due to Kato [37]:

THEOREM 1.3.17 (Kato). Suppose that B is simple normally crossing. Let K be a normal subgroup of $\pi_1(M-B, p_o)$ of finite index such that $J \subset K$. Suppose that K satisfies Condition 1.3.3 and the following condition (under the notations in the proof of Theorem 1.3.15):

$$\text{if } \hat{\gamma}_k^{d_k} \ldots \hat{\gamma}_n^{d_n} \in K, \text{ then } d_k \equiv 0 \pmod{e_k}, \ldots, d_n \equiv 0 \pmod{e_n}$$

for every point $p \in \mathrm{Sing}B$. Then the finite Galois covering $\pi : X \longrightarrow M$ which branches at D, corresponding to K by Theorem 1.3.9, is of simple normally crossing type.

Note that the last condition is equivalent to

$$K_p = \mathbb{Z}(e_k\hat{\gamma}_k) + \ldots + \mathbb{Z}(e_n\hat{\gamma}_n).$$

The proof of the theorem is hence similar to that of Corollary 1.3.16.

Now we give some examples which explain the above discussion.

1°. Put $M = \mathbb{C}^2$ and

$$B = \{(v,w) \in \mathbb{C}^2 \mid w^2 = v^3\}.$$

Then B is an irreducible curve in M. Put

$$W = \{(v,w) \in \mathbb{C}^2 \mid |v|^2 + |w|^2 < 1\}.$$

Then for a fixed point $p_0 \in W-B$, the inclusion map $i : W-B \hookrightarrow M-B$ induces the isomorphism

$$i_* : \pi_1(W-B,\ p_0) \simeq \pi_1(M-B,\ p_0).$$

We identify these groups through i_*. It is well known (see Milnor [41]) that $(\bar{W}, \bar{W} \cap B)$ is homeomorphic to the cone $\mathrm{Cone}(\partial\bar{W}, \partial\bar{W} \cap B)$. (Here, $\bar{W}$ is the closure of W and $\partial\bar{W}$ is its boundary.)

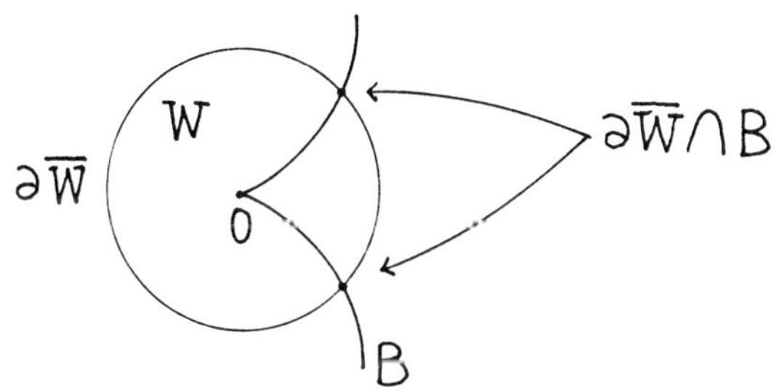

Hence

$$\pi_1(W-B,\ p_0) \cong \pi_1(\partial\bar{W}-B,\ p_0) \cong \pi_1(S^3-C,\ q_0),$$

where S^3 is the unit 3-sphere: $S^3 = \{(v,w) \in \mathbb{C}^2 \mid |v|^2 + |w^2| = 1\}$ and C is a, so-called, trefoil knot in S^3.

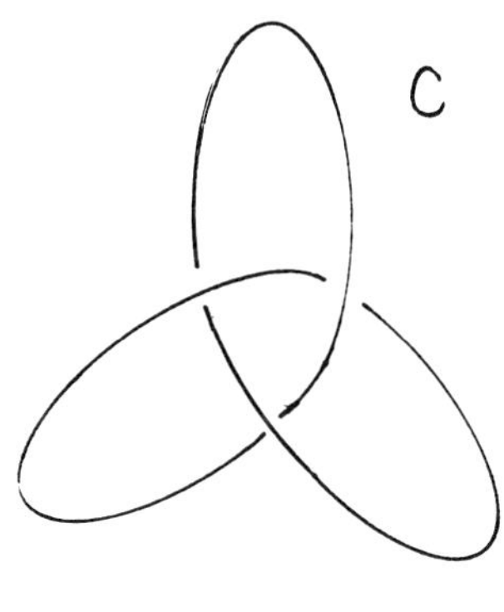

Again, we identify these groups through the isomorphisms. It is well known (see Fox-Crowell [16]) that $\pi_1(S^3-C, q_o)$ is generated by three elements α, β and γ with the following relations:

$$\pi_1(S^3-C, q_o) = \langle\alpha,\beta,\gamma \mid \alpha\beta\alpha^{-1} = \gamma,\ \beta\gamma\beta^{-1} = \alpha,\ \gamma\alpha\gamma^{-1} = \beta\rangle .$$

Eliminating γ, we may also write

$$\pi_1(S^3-C, q_o) = \langle\alpha,\beta \mid \alpha\beta\alpha = \beta\alpha\beta\rangle.$$

Let γ_1 be as before a closed path in M-B which starts from p_o and rounds B once in a counterclockwise direction. Then α,β,γ and γ_1 are conjugate in $\pi_1(M-B, p_o)$ to each other. Hence we may put $\gamma_1 = \alpha$, say.

Let e be a positive integer and put D = eB. Put

$$X = \{(u,v,w) \in \mathbb{C}^3 \mid u^e + v^3 - w^2 = 0\},$$

$$\pi : (u,v,w) \in X \longmapsto (v,w) \in M = \mathbb{C}^2 .$$

Then X is a normal surface in $\mathbb{C}^3$ and π is a cyclic covering which branches at D. hence, by Lemma 1.3.2, $K = \pi_*(\pi_1(X-\pi^{-1}(B), r_o))$ satisfies Condition 1.3.3. The smallest normal subgroup J of $\pi_1(M-B, p_o)$ which contains α^e is contained in K and also satisfies Condition 1.3.3.

Note that K is of finite index in $\pi_1(M-B, p_o)$. In the present case, this is equivalent to Condition 1.3.6 for K. Let $\tilde{K}$ be the intersection of all subgroups K' of $\pi_1(M-B, p_o)$ of finite index such that $J \subset K'$.

PROBLEM. When is $\tilde{K}$ (or J) of finite index in $\pi_1(M-B, p_o)$?

We do not know the answer. We give here a partial answer.

LEMMA 1.3.18 Under the above notations, $G = \pi_1(M-B, p_o)/J$ is the group generated by two elements a and b with the following relations:

$$G = \langle a,b \mid a^e = b^e = 1 \ \text{ and } \ aba = bab\rangle.$$

PROOF. Let F be the free group generated by two elements $\hat{\alpha}$ and $\hat{\beta}$: $F = \langle\hat{\alpha},\hat{\beta}\rangle$. Let N (resp. P) be the smallest normal subgroup of F which contains $\hat{\alpha}\hat{\beta}\hat{\alpha}\hat{\beta}^{-1}\hat{\alpha}^{-1}\hat{\beta}^{-1}$ (resp., $\hat{\alpha}\hat{\beta}\hat{\alpha}\hat{\beta}^{-1}\hat{\alpha}^{-1}\hat{\beta}^{-1}, \hat{\alpha}^e$ and $\hat{\beta}^e$). We define a surjective homomorphism

$$f : F \longrightarrow \pi_1(M-B, p_o)$$

by $f(\hat{\alpha}) = \alpha$ and $f(\hat{\beta}) = \beta$. Then

$$\ker(f) = N \ \text{ and so, } \ f : F/N \simeq \pi_1(M-B, p_o).$$

Note that the image of P by f is J. In fact, since f(P) is a normal subgroup of $\pi_1(M-B, p_o)$ and $\alpha^e = f(\hat{\alpha}^e) \in f(P)$, we have $J \subset f(P)$. On the other hand, we can easily show that P is generated by the elements of the form

$$\hat{\delta}^{-1}(\hat{\alpha}\hat{\beta}\hat{\alpha}\hat{\beta}^{-1}\hat{\alpha}^{-1}\hat{\beta}^{-1})\hat{\delta}, \quad \hat{\delta}^{-1}(\hat{\alpha}^e)\hat{\delta} \quad \text{and} \quad \hat{\delta}^{-1}(\hat{\beta}^e)\hat{\delta},$$

where $\hat{\delta}$ moves arbitrary element of F. We have

$$f(\hat{\delta}^{-1}(\hat{\alpha}\hat{\beta}\hat{\alpha}\hat{\beta}^{-1}\hat{\alpha}^{-1}\hat{\beta}^{-1})\hat{\delta}) = \delta^{-1}\alpha\beta\alpha\beta^{-1}\alpha^{-1}\beta^{-1}\delta = \delta^{-1}\delta = 1,$$

$$f(\hat{\delta}^{-1}(\hat{\alpha}^e)\hat{\delta}) = \delta^{-1}\alpha^e\delta \in J,$$

$$f(\hat{\delta}^{-1}(\hat{\beta}^e)\hat{\delta}) = \delta^{-1}\beta^e\delta \in J. \qquad (\delta = f(\delta).)$$

Hence $f(P) \subset J$. Thus $f(P) = J$. Hence

$$\pi_1(M-B, p_o)/J \simeq (F/N)/(P/N) \simeq F/P.$$

But

$$F/P \simeq \langle a,b \mid a^e = b^e = 1,\ aba = bab\rangle. \qquad \square$$

Consider the homomorphism $h : G \longrightarrow \langle c \mid c^e = 1\rangle \ (\simeq \mathbb{Z}/e\mathbb{Z})$ defined by

$$h(a) = h(b) = c.$$

Then h is well defined and is surjective. Its kernel H = ker(h) corresponds to $K = \pi_*(\pi_1(X-\pi^{-1}(B),\ r_o))$, that is, $H = K/J$, where $\pi : X \longrightarrow M = \mathbb{C}^2$ is the above cyclic covering. The order of a (and b) is e. In fact, if $a^d = 1$ for $1 \leq d < e$, then $1 = f(a^d) = c^d$, a contradiction. (This implies again that K satisfies Condition 1.3.3.)

Now we discuss the structure of G.

(i) If e = 2, then

$$G = \langle a,b \mid a^2 = b^2 = 1,\ aba = bab\rangle.$$

Then

$$(ab)^3 = ababab = abaaba = 1.$$

Conversely, if $a^2 = b^2 = 1$ and $(ab)^3 = 1$, then

$$aba = b^{-1}a^{-1}b^{-1} = bab.$$

Hence

$$G = \langle a,b \mid a^2 = b^2 = (ab)^3 = 1\rangle.$$

This is the dihedral group D_3 (see Theorem 1.2.17), which is isomorphic to 3-rd symmetric group S_3.

Thus, by Theorem 1.3.11, there is a maximal covering

$$\tilde{\pi} : \tilde{M}(D) \longrightarrow M = \mathbb{C}^2$$

which branches at $D = 2B$ such that $G_{\tilde{\pi}} \simeq S_3$. (This is nothing but the covering $\pi : X \longrightarrow M$ in Example 1.1.18.)

(ii). If $e = 3$, then

$$G = \langle a,b \mid a^3 = b^3 = 1, \quad aba = bab\rangle.$$

Note that

$$(ab)^2 = abab = aaba = a^{-1}ba.$$

Hence the order of $(ab)^2$ is 3 and $(ab)^6 = 1$. Hence the order of ab is either 3 or 6. We show that the order of ab is 6.

Using the relations $a^{-1} = a^2$ and $b^{-1} = b^2$, we can write every element of G as a finite product of a and b. Then we can easily show that $H = \ker(h)$ consists of products of 3k ($k = 0,1,2,\ldots$) numbers of a and b. Thus H is generated by the following elements:

$$ba^2,\ a^2b,\ aba,\ ab^2,\ b^2a,\ bab.$$

But note that

$$ab^2 = (ba^2)^{-1}, \quad b^2a = (a^2b)^{-1} \quad \text{and} \quad bab = aba.$$

Hence H is generated by

$$p = ba^2, \quad q = a^2b \quad \text{and} \quad r = aba,$$

which satisfy the following relations:

$$p^2 = q^2 = r^2 = s, \quad s^2 = 1,$$

$$pq = r, \quad qr = p, \quad rp = q,$$

$$qp = r^{-1} = sr = rs,$$

$$rq = p^{-1} = sp = ps,$$

$$pr = q^{-1} = sq = qs.$$

These relations remind us of the subgroup Q_o of all units Q^* of the (noncommutative) field Q of quaternions:

$$Q_o = \{1, i, j, k, -1, -i, -j, -k\}.$$

By defining $g(i) = p$, $g(j) = q$, $g(k) = r$ and $g(-1) = s$, we have a surjective homomorphism

$$g : Q_o \longrightarrow H.$$

We show that g is in fact an isomorphism. Consider the permutation

$$\tau = \begin{pmatrix} 1 & i & j & k & -1 & -i & -j & -k \\ 1 & j & k & i & -1 & -j & -k & -i \end{pmatrix}$$

of the elements of Q_o. Then τ is an automorphism of the group Q_o which is not an inner automorphism. Note also that $\tau^3 = 1$.

Consider the set

$$\hat{G} = \{c^\nu.x \mid 0 \leq \nu \leq 2,\ x \in Q_o\}$$

of all formal products of $c^\nu \in \langle c \mid c^3 = 1\rangle$ and $x \in Q_o$. We define the product of two elements of $\hat{G}$ as follows:

$$(c^\nu.x)(c^\mu.y) = c^{\nu+\mu}.\tau^\mu(x)y.$$

Then $\hat{G}$ becomes a group. The map

$$c^\nu.x \in \hat{G} \longmapsto c^\nu \in \langle c \mid c^3 = 1\rangle$$

is a surjective homomorphism with the kernel Q_o. Hence

$$1 \longrightarrow Q_o \longrightarrow \hat{G} \longrightarrow \langle c \mid c^3 = 1\rangle \longrightarrow 1$$

is exact.

We have a surjective homomorphism $G \longrightarrow \hat{G}$ by sending $a \longrightarrow c.1$ and $b \longrightarrow c.j$. Then $ab \longrightarrow c^2.j$. Note that $c^2.j$ has the order 6. This implies that the order of ab is 6. Since

$$1 \longrightarrow H \longrightarrow G \longrightarrow \langle c \mid c^3 = 1\rangle \longrightarrow 1$$

is exact and $g : Q_o \longrightarrow H$ is surjective, the order of G is a divisor of 24. Hence $G \longrightarrow \hat{G}$ is in fact an isomorphism and so the order of G is 24. Thus $g : Q_o \longrightarrow H$ is an isomorphism and so there is the following exact sequence:

$$1 \longrightarrow Q_o \longrightarrow G \longrightarrow \langle c \mid c^3 = 1\rangle \longrightarrow 1.$$

Now, by Theorem 1.3.11, there is a maximal covering

$$\tilde{\pi} : \tilde{M} = \tilde{M}(D) \longrightarrow M = \mathbb{C}^2$$

which branches at $D = 3B$ such that $G_{\tilde{\pi}} = G$. Also, there is the following diagram:

$$\begin{array}{ccl} \tilde{M} & \xrightarrow{\ \mu\ } & X = \{(u,v,w) \in \mathbb{C}^3 \mid u^3 + v^3 - w^2 = 0\}. \\ {\scriptstyle \tilde{\pi}} \downarrow & \swarrow {\scriptstyle \pi} & \\ M & & \end{array}$$

Since π branches at D,

$$\mu : \tilde{M} - \mu^{-1}(0) = \tilde{M} - \tilde{\pi}^{-1}(0) \longrightarrow X - \{0\}$$

is unramified. By Theorem 1.3.11, $\tilde{M} - \tilde{\pi}^{-1}(0)$ is simply connected. Thus we have

PROPOSITION 1.3.19 Put $X = \{(u,v,w) \in \mathbb{C}^3 \mid u^3 + v^3 - w^2 = 0\}$. Then $\pi_1(X - \{0\}, r_0)$ is isomorphic to the group

$$Q_0 = \{1,i,j,k,-1,-i,-j,-k\}.$$

(iii) As for the structure of G for a general e, we sketch the discussion with Mr. Mizutani. (See also Coxeter [6].) Put

$$z = (aba)^2.$$

LEMMA 1.3.20 $z = (aba)^2$ is an element of the centre $Z(G)$ of G.

PROOF. Note that

$$z = (aba)^2 = (ab)^3 = (bab)^2 = (ba)^3.$$

Hence

$$z(aba) = (aba)z,$$

$$z(ab) = (ab)z.$$

Hence

$$za = az.$$

In a similar way,

$$zb = bz.$$

Hence z is an element of the center $Z(G)$ of G. □

Consider the *Schwarz triangular group*:

$$G(2,3,e) = \langle S,T \mid S^2 = T^3 = (ST)^e = 1\rangle.$$

LEMMA 1.3.21 There is the following exact sequence

$$1 \longrightarrow \langle z\rangle \longrightarrow G \xrightarrow{f} G(2,3,e) \longrightarrow 1$$

where $\langle z\rangle$ is the cyclic group generated by z and the homomorphism f is defined by $f(a) = ST$ and $f(b) = TS$.

PROOF. f is well defined, because

$$(ST)^e = (TS)^e = 1,$$

$$(ST)(TS)(ST) = S = (TS)(ST)(TS).$$

f is surjective, because

$$f(aba) = S \quad \text{and} \quad f((ba)^{-1}) = T.$$

Let $\bar{a}$ and $\bar{b}$ be the images of a and b, respectively, of the projection

$$p : G \longrightarrow G/\langle z\rangle.$$

Consider the homomorphism

$$g : G(2,3,e) \longrightarrow G/\langle z\rangle$$

defined by $g(S) = aba$ and $g(T) = (ba)^{-1}$. (g is then well defined.) g is surjective, because

$$g(ST) = a \quad \text{and} \quad g(TS) = b.$$

Note that

$$gf(a) = a \quad \text{and} \quad gf(b) = b.$$

hence

$$gf = p.$$

Hence

$$\mathrm{Ker}(f) \subset \langle z\rangle.$$

On the other hand, z is an element of Ker(f), because

$$f(z) = f(aba)^2 = S^2 = 1.$$

Thus

$$\langle z\rangle = \mathrm{Ker}(f). \quad \square$$

We know the structure of $G(2,3,e)$ for small e, (see, for example, Coxeter-Moser [7]). By the above lemma, we know the structure of G small e:

e	ord(z)	G(2,3,e)	G	ord(G)
2	1	S_3	S_3	6
3	2	A_4	$SL(2, \mathbb{Z}/3\mathbb{Z})$	24
4	4	S_4	$G/Z(G) \simeq S_4$ $Z(G) \simeq \mathbb{Z}/4\mathbb{Z}$	96
5	10	A_5	$SL(2, \mathbb{Z}/5\mathbb{Z}) \times (\mathbb{Z}/5\mathbb{Z})$	600
6	∞	infinite group	infinite solvable group	∞
$e \geq 7$	∞	infinite group	infinite unsolvable group	∞

(Here, S_n and A_n are symmetric and alternating groups of n letters, respectively. ord(z) and ord(G) are the orders of z and G, respectively.)

Thus, if $e \leq 5$, then, by Theorem 1.3.11, there is a maximal covering

$$\tilde{\pi} : \tilde{M}(D) \longrightarrow M = \mathbb{C}^2$$

which branches at $D = eD_1$, $(e \leq 5)$. $\tilde{\pi}$ is a finite Galois covering in this case. M(D) can be identified with $\mathbb{C}^2$ and $\tilde{\pi}$ splits into

$$\tilde{\pi} : \tilde{M}(D) = \mathbb{C}^2 \xrightarrow{\mu} X \xrightarrow{\pi} M = \mathbb{C}^2,$$

where π and X are defined as before:

$$X = \{(u,v,w) \in \mathbb{C}^3 \mid u^e + v^3 - w^2 = 0\}, \quad (e \leq 5)$$

$$\pi : (u,v,w) \in X \longrightarrow (v,w) \in \mathbb{C}^2,$$

and μ is the projection

$$\mu : \tilde{M}(D) = \mathbb{C}^2 \longrightarrow X = \mathbb{C}^2/H.$$

Here H is a finite subgroup of $GL(2, \mathbb{C})$. The origin of X is called the *Klein singularity*, (see Pinkham [49]).

(iv) If $e = 6$, then G is an infinite solvable group. In this case, we have the following two lemmas, whose proofs are omitted:

LEMMA 1.3.22 The kernel N of the homomorphism

$$h : G \longrightarrow \langle c \mid c^6 = 1 \rangle \quad (\simeq \mathbb{Z}/6\mathbb{Z})$$

defined by $h(a) = h(b) = c$, is generated by $a^{-1}b$ and ab^{-1}, and is isomorphic to the group

$$N' = \left\{ \begin{pmatrix} 1 & k & \ell \\ 0 & 1 & m \\ 0 & 0 & 1 \end{pmatrix} \middle| \ k, \ell, m \in \mathbb{Z} \right\}$$

by the mapping

$$a^{-1}b \mapsto \begin{pmatrix} 1 & 1 & 0 \\ 0 & 1 & 0 \\ 0 & 0 & 1 \end{pmatrix} \quad ab^{-1} \mapsto \begin{pmatrix} 1 & 0 & 0 \\ 0 & 1 & 1 \\ 0 & 0 & 1 \end{pmatrix}.$$

LEMMA 1.3.23 $G = HN$ (a semi-direct product), where H is the cyclic group generated by a.

We identify N with N' by Lemma 1.3.22. Then we have

$$a^{-1} \begin{pmatrix} 1 & k & \ell \\ 0 & 1 & m \\ 0 & 0 & 1 \end{pmatrix} a = \begin{pmatrix} 1 & k-m & \ell-km+k(k+1)/2 \\ 0 & 1 & k \\ 0 & 0 & 1 \end{pmatrix} \qquad \ldots(*)$$

For any odd positive integer r, put

$$N(r) = \left\{ \begin{pmatrix} 1 & k & \ell \\ 0 & 1 & m \\ 0 & 0 & 1 \end{pmatrix} \middle| \ k \equiv \ell \equiv m \equiv 0 \pmod r \right\}.$$

Then, by (*), $N(r)$ is a normal subgroup of G of index $6r^3$. Moreover, the pullback of $N(r)$ to $\pi_1(\mathbb{C}^2 - B, p_o)$ satisfies Condition 1.3.3. Thus, by Theorem 1.3.9,

PROPOSITION 1.3.24 For any positive odd integer r, there is a Galois covering $\nu_r : Y_r \longrightarrow M = \mathbb{C}^2$ of degree $6r^3$ which branches at $6D_1$. $\nu_r \leq \nu_{r'}$ if and only if r divides r'.

Note that

$$\bigcap_r N(r) = \{1\},$$

where $\cap$ runs over all odd positive integers r. Thus, by Theorem 1.3.10,

PROPOSITION 1.3.25 There *does not* exist a maximal covering $\tilde{\pi} : \tilde{M} \longrightarrow M = \mathbb{C}^2$ which branches at $6D_1$.

2°. Next, consider the case

$M = \mathbb{P}^2$, the complex projective plane,

D_1 = the closure in $\mathbb{P}^2$ of $\{(v,w) \in \mathbb{C}^2 \mid v^3 - w^2 = 0\}$,

$D_2 = L_\infty$ (the line of infinity),

$B = D_1 \cup D_2$,

e, m : positive integers greater than 1,

$D = ed_1 + mD_2$,

where (v,w) is an affine coordinate in $\mathbb{P}^2$.

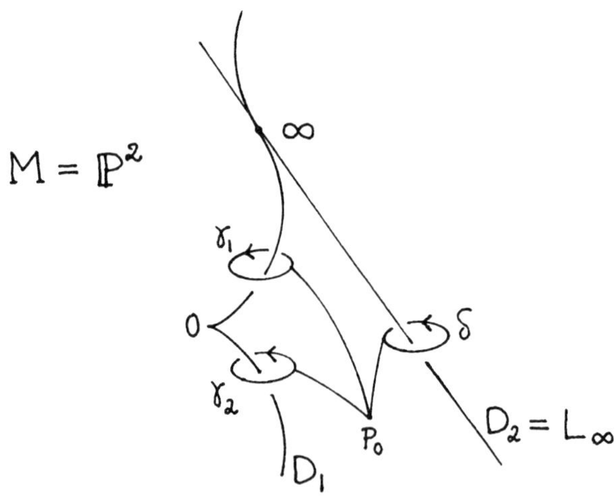

We choose a closed path δ as in the figure. Then, as is easily seen,

$$\gamma_1\gamma_2\gamma_1 = \gamma_2\gamma_1\gamma_2 = \delta^{-1}.$$

Let J be the smallest normal subgroup of $\pi_1(\mathbb{P}^2\text{-}B, p_0)$ containing γ_1^e, γ_2^e and δ^m. Then we have

$$\pi_1(\mathbb{P}^2 - B, p_o)/J \simeq G_{e,m},$$

where

$$G_{e,m} = \langle \alpha, \beta, \delta \mid \alpha^e = \beta^e = \delta^m = 1, \quad \alpha\beta\alpha = \beta\alpha\beta = \delta^{-1} \rangle .$$

The isomorphism is given by

$$\gamma_1 \longmapsto \alpha, \quad \gamma_2 \longmapsto \beta, \quad \delta \longmapsto \delta.$$

We identify $\pi_1(\mathbb{P}^2 - B, p_o)/J$ with $G_{e,m}$. Let

$$G_e = \langle a, b \mid a^e = b^e = 1, \quad aba = bab \rangle$$

be the group in 1°. We then have a surjective homomorphism

$$f : G_e \longrightarrow G_{e,m}$$

by defining $f(a) = \alpha$ and $f(b) = \beta$.

For simplicity, *we assume that* m *is even.* Then we have easily,

LEMMA 1.3.26 The kernel of f is $\langle z^{m/2} \rangle$, where $z = (aba)^2$, (see Lemma 1.3.20).

Hence we have the following exact sequence:

$$1 \longrightarrow \langle z^{m/2} \rangle \longrightarrow G_e \xrightarrow{f} G_{e,m} \longrightarrow 1.$$

In particular, if the pair (e,m) is one of (2,2), (3,4), (4,8) and (5,20), then $G_{e,m} \simeq G_e$. Hence

PROPOSITION 1.3.27 If the pair (e,m) is one of (2.2), (3,4), (4,8) and (5,20), then there is a maximal covering $\tilde{\pi} : \tilde{M}(D) \longrightarrow M = \mathbb{P}^2$ which branches at $D = eD_1 + mD_2$. In this case, $\tilde{\pi}$ is a finite covering such that $G_{\tilde{\pi}} \simeq G_{e,m} \simeq G_e$.

If e = 6, then, by Proposition 1.3.24,

PROPOSITION 1.3.28 For any positive integer m such that $m \equiv 2 \pmod 4$, there is a Galois covering $\rho_m : Z_m \longrightarrow M = \mathbb{P}^2$ of degree $6(m/2)^3$ which branches at $D = 6D_1 + mD_2$. $\rho_m \leq \rho_{m'}$ if and only if m divides m'.

Combining the above exact sequence and the exact sequence in 1°, we have the following exact sequence:

$$1 \longrightarrow \langle z \rangle / \langle z^{m/2} \rangle \longrightarrow G_{e,m} \longrightarrow G(2,3,e) \longrightarrow 1.$$

In particular, if $m = 2$, then

$$G_{e,2} \simeq G(2,3,e).$$

Taking $e = 6$,

$$G_{6,2} \simeq G(2,3,6).$$

We identify $G_{6,2}$ with $G(2,3,6)$ through this isomorphism. It is well known that there is a normal subgroup L of $G(2,3,6) = G_{6,2}$ such that

$$1 \longrightarrow L \longrightarrow G(2,3,6) \longrightarrow \mathbb{Z}/6\mathbb{Z} \longrightarrow 1,$$

$$L \simeq \mathbb{Z} \oplus \mathbb{Z} \quad \text{(a direct sum).}$$

We identify L with $\mathbb{Z} \oplus \mathbb{Z}$. For any positive integer q, put

$$L(q) = \{(j,k) \in \mathbb{Z} \oplus \mathbb{Z} \mid j \equiv k \equiv 0 \pmod{q}\}.$$

Then $L(q)$ is a normal subgroup of $G(2,3,6) = G_{6,2}$ of index $6q^2$. Note that

$$\bigcap_q L(q) = 1.$$

Hence

PROPOSITION 1.3.29 There *does not* exist a maximal covering $\tilde{\pi} : \tilde{M}(D) \longrightarrow M = \mathbb{P}^2$ which branches at $D = 6D_1 + 2D_2$.

3°. Next, put

$$M = \mathbb{P}^2,$$

$$B = \{(X_0:X_1:X_2) \in \mathbb{P}^2 \mid X_0X_2^2 - X_1^3 = 0\}.$$

(Here $(X_0:X_1:X_2)$ is a homogeneous coordinate system in $\mathbb{P}^2$.)

Then B is the closure in $\mathbb{P}^2$ of the irreducible affine curve $\{(v,w) \in \mathbb{C}^2 \mid v^3 - w^2 = 0\}$. Following Zeriski [69], we determine $\pi_1(\mathbb{P}^2 - B, p_o)$.

Let $p_o \in \mathbb{P}^2 - B$ be a fixed point and ℓ be a general line passing through p_o and cutting B at three distinct points p_1, p_2 and p_3.

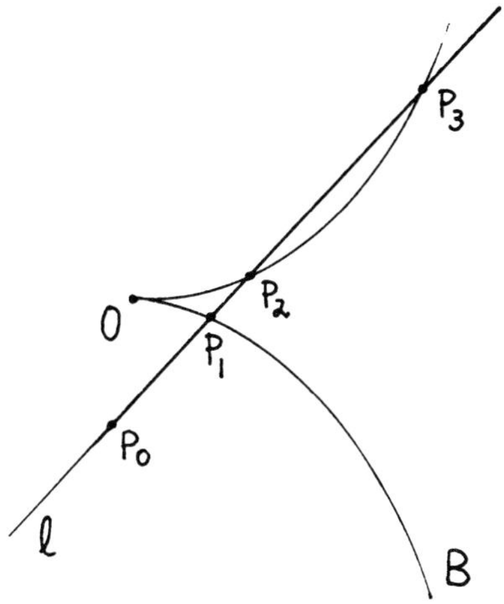

It can be shown that any closed path in $\mathbb{P}^2$-B which starts from p_o is homotopic to a closed path in $\ell - \{p_1, p_2, p_3\}$ which starts from p_o. Let β_j $(1 \leq j \leq 3)$ be a closed path in $\ell - \{p_1, p_2, p_3\}$ which starts from p_o and rounds p_j once in a counterclockwise direction as in the figure:

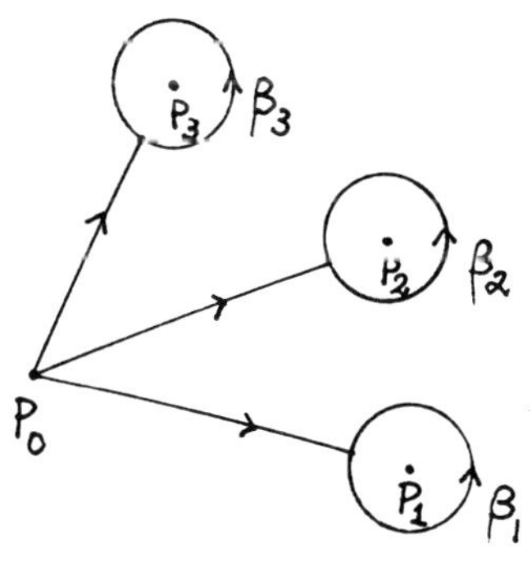

Note that

$$\pi_1(\ell - \{p_1, p_2, p_3\},\ p_o) = \langle \beta_1, \beta_2, \beta_3 \mid \beta_1\beta_2\beta_3 = 1 \rangle .$$

We regard β_1, β_2 and β_3 as elements of $\pi_1(\mathbb{P}^2\text{-B},\ p_o)$. Then they generate $\pi_1(\mathbb{P}^2\text{-B},\ p_o)$ and satisfy the following conditions:

$$\beta_1\beta_2\beta_3 = 1, \quad \beta_1\beta_2\beta_1^{-1} = \beta_3, \quad \beta_2\beta_3\beta_2^{-1} = \beta_1, \quad \beta_3\beta_1\beta_3^{-1} = \beta_2,$$

(see 1° above). From these relations, we have

$$\beta_1 = \beta_2 = \beta_3, \ \beta_1^3 = 1.$$

Hence

$$\pi_1(\mathbb{P}^2 - B, p_o) = \langle \beta_1 \mid \beta_1^3 = 1 \rangle \simeq \mathbb{Z}/3\mathbb{Z}.$$

This implies that the cyclic covering

$$\begin{array}{ccc} \pi : X = \{(X_0:X_1:X_2:X_3) \in \mathbb{P}^3 \mid X_2^3 - X_0X_2^2 + X_1^3 = 0\} & \longrightarrow & \mathbb{P}^2 \\ \cup & & \cup \\ (X_0:X_1:X_2:X_3) & \longmapsto & (X_0:X_1:X_2) \end{array}$$

is a maximal covering which branches at $D = 3B$. In particular, X and $X - \{(1:0:0:0)\}$ are simply connected.

More generally, Zariski [69] proved the following theorem:

THEOREM 1.3.30 (Zariski). If B is an irreducible cubic or quartic curve in $\mathbb{P}^2$, then $\pi_1(\mathbb{P}^2 - B, p_o)$ is cyclic of order 3 and 4, respectively, unless B is a 3-cuspidal quartic. If B is a 3-cuspidal quartic, then

$$\pi_1(\mathbb{P}^2 - B, p_o) \simeq \langle \beta_1, \beta_2 \mid \beta_1^2 = \beta_2^2, \ \beta_1^4 = 1, \ (\beta_1\beta_2)^3 = \beta_1^2 \rangle,$$

which has the order 12 and is not cyclic.

It is well known that 3-cuspidal quartic curves in $\mathbb{P}^2$ are rational curves and projectively equivalent to each other. Put

$$F(X_0, X_1, X_2) = (X_1^2 + X_2^2 - X_0X_1)^2 - X_0^2X_1^2 - X_0^2X_2^2,$$

$$B = \{(X_0:X_1:X_2) \in \mathbb{P}^2 \mid F(X_0, X_1, X_2) = 0\}.$$

Then B is such an example. The 3 simple cusps of B are

$$p_1 = (1:0:0), \quad p_2 = (0:1:\sqrt{-1}) \quad \text{and} \quad p_3 = (0:1:-\sqrt{-1}),$$

(see, for example, Namba [44]).

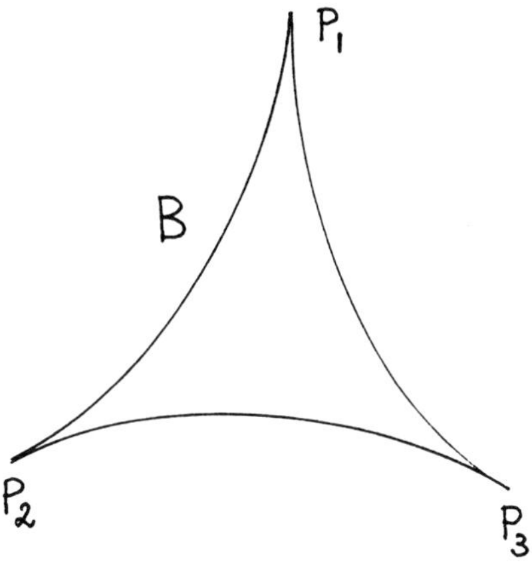

An obvious covering of $\mathbf{P}^2$ which branches at $D = 4B$ is the cyclic covering

$$\pi : (X_0:X_1:X_2:X_3) \in X \longmapsto (X_0:X_1:X_2) \in \mathbf{P}^2,$$

where

$$X = \{(X_0:X_1:X_2:X_3) \in \mathbf{P}^3 \mid X_3^4 - F(X_0,X_1,X_2) = 0\}.$$

X is a normal projective surface with the 3 singular points

$$q_1 - (1:0:0:0), \quad q_2 - (0:1:\sqrt{-1}:0) \quad \text{and} \quad q_3 - (0:1:-\sqrt{-1}:0),$$

each of which is locally given by the equation:

$$x^2 + y^3 + z^4 = 0.$$

The above theorem of Zariski asserts that there is a maximal covering

$$\tilde{\pi} : \widetilde{\mathbf{P}^2}(D) \longrightarrow \mathbf{P}^2$$

which branches at $D = 4B$ such that

$$G_{\tilde{\pi}} \simeq \pi_1(\mathbf{P}^2 - B, p_o) \simeq \langle \beta_1, \beta_2 \mid \beta_1^2 = \beta_2^2,\ \beta_1^4 = 1,\ (\beta_1\beta_2)^3 = \beta_1^2 \rangle.$$

There is a holomorphic map $\mu : \widetilde{\mathbf{P}^2}(D) \longrightarrow X$ such that $\pi\mu = \tilde{\pi}$. Hence, as in Proposition 1.3.1,

COROLLARY 1.3.31 Let X be the normal projective surface defined above. Then $\pi_1(X - \{q_1,q_2,q_3\}, r_o) \simeq \mathbb{Z}/3\mathbb{Z}$.

4°. As the final example, put $M = \mathbf{P}^2$ and

$$B = D_1 \cup \ldots \cup D_s$$

is a nodal curve (i.e. it has only nodes as its singularity), where each D_j is an irreducible plane curve of degree d_j.

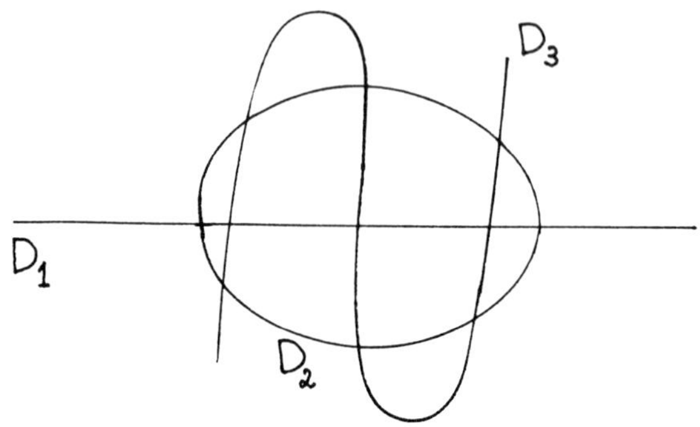

Let γ_j be, as before, a closed path in $\mathbb{P}^2$-B which starts from a fixed point p_o and rounds D_j-SingB once in a counterclockwise direction. Since $\pi_1(\mathbb{P}^2\text{-SingB}, p_o) \simeq \pi_1(\mathbb{P}^2, p_o) = \{1\}$, $\pi_1(\mathbb{P}^2\text{-B}, p_o)$ is itself equal to the smallest normal subgroup J_o which contains $\gamma_1, \ldots, \gamma_s$, by Corollary 1.2.8. We recall the celebrated

THEOREM 1.3.32 (Zariski [69], Fulton [18], Deligne [9]). $\pi_1(\mathbb{P}^2\text{-B}, p_o)$ is the *abelian* group, generated by $\gamma_1, \ldots, \gamma_s$ with the unique relation $\gamma_1^{d_1} \cdots \gamma_s^{d_s} = 1$.

Thus, we may use addition instead of multiplication as group operation. We may write

$$\pi_1(\mathbb{P}^2\text{-B}, p_o) = \mathbb{Z}\gamma_1 + \ldots + \mathbb{Z}\gamma_s,$$

where $d_1\gamma_1 + \ldots + d_s\gamma_s = 0$. Let

$$F = \mathbb{Z}\hat{\gamma}_1 + \ldots + \mathbb{Z}\hat{\gamma}_s$$

be the free additive group with the basis $\{\hat{\gamma}_1, \ldots, \hat{\gamma}_s\}$. Consider the subgroup

$$N = \mathbb{Z}(d_1\hat{\gamma}_1 + \ldots + d_s\hat{\gamma}_s),$$

of F. Then

$$\begin{array}{ccccccccc} 0 & \longrightarrow & N & \longrightarrow & F & \xrightarrow{f} & \pi_1(\mathbb{P}^2\text{-B}, p_o) & \longrightarrow & 0 \\ & & & & \cup\!\!\!\!\!\!\;\; & & \cup & & \\ & & & & \hat{\gamma}_j & \longmapsto & \gamma_j & & \end{array}$$

is exact. Let J be the subgroup of $\pi_1(\mathbb{P}^2\text{-B}, p_o)$ generated by $e_1\gamma_1, \ldots, e_s\gamma_s$:

$$J = \mathbb{Z}(e_1\gamma_1) + .. + \mathbb{Z}(e_s\gamma_s).$$

Then

$$\hat{J} = f^{-1}(J) = N + \mathbb{Z}(e_1\hat{\gamma}_1) + \ldots + \mathbb{Z}(e_s\hat{\gamma}_s).$$

Put

$$G = \pi_1(\mathbb{P}^2 - B,\ p_o)/J.$$

Then

$$G \simeq (F/N)/(\hat{J}/N) \simeq F/\hat{J}.$$

Hence we may write

$$G = \mathbb{Z}\gamma_1 + \ldots + \mathbb{Z}\gamma_s,$$

where $d_1\gamma_1 + \ldots + d_s\gamma_s = 0$ and $e_j\gamma_j = 0$ for $1 \leq j \leq s$. (Here, by abuse of notation, we denote also by γ_j the image of γ_j under the projection $\pi_1(\mathbb{P}^2\text{-}B,\ p_o) \longrightarrow G$.) In particular, G is a finite abelian group.

<u>LEMMA 1.3.33</u>. J satisfies Condition 1.3.3 if and only if, for any $j(1 \leq j \leq s)$, $e_j/(d_j,e_j)$ divides

$$\langle e_1/(d_1,e_1),\ldots,\check{e_j/(d_j,e_j)},\ldots,e_s/(d_s,e_s)\rangle,$$

where (d_j,e_j) denotes the GCD of d_j and e_j and $\langle a_1,\ldots,\check{a}_j,\ldots,a_s\rangle$ denotes the LCM of $a_1,\ldots,a_s$ except a_j. (For convenience, put $\langle a_1,\ldots,\check{a}_j,\ldots,a_s\rangle = 1$ if $s = 1$.)

<u>PROOF</u>. For an integer d, suppose that $d\hat{\gamma}_1 \in \hat{J}$. Then

$$d\hat{\gamma}_1 = k_0(d_1\hat{\gamma}_1 + \ldots + d_s\hat{\gamma}_s) + k_1e_1\hat{\gamma}_1 + \ldots + k_se_s\hat{\gamma}_s$$

for some integers $k_0,k_1,\ldots,k_s$. Then

$$d = k_0d_1 + k_1e_1,$$

$$0 = k_0d_2 + k_2e_2,$$

$$\ldots\ldots$$

$$0 = k_0d_s + k_se_s.$$

Hence

$$k_0 \equiv 0 \pmod{m_1},$$

where

$$m_1 = \langle e_2/(d_2,e_2),\dots,e_s/(d_s,e_s)\rangle.$$

Put $k_0 = am_1$. Then

$$d = am_1d_1 + k_1e_1.$$

Thus

$$d \equiv 0 \pmod{(m_1d_1,e_1)}. \tag{1}$$

Conversely, if $d = am_1d_1 + k_1e_1$ for some integers a and k_1, then, by putting

$$k_0 = am_1,\quad k_2 = -k_0d_2/e_2,\dots,\ k_s = -k_0d_s/e_s,$$

we have

$$d\hat{\gamma}_1 = k_0(d_1\hat{\gamma}_1 + \dots + d_s\hat{\gamma}_s) + k_1e_1\hat{\gamma}_1 + \dots + k_se_s\hat{\gamma}_s.$$

Thus, if Condition 1.3.3 is satisfied, then, by (1),

$$(m_1d_1,e_1) = e_1,$$

that is

$$e_1/(d_1,e_1) \ \text{ divides } m_1.$$

In a similar way,

$$e_j/(d_j,e_j) \ \text{ divides } m_j,$$

where

$$m_j = \langle e_1/(d_1,e_1),\dots,\check{e_j/(d_j,e_j)},\dots,e_s/(d_s,e_s)\rangle.$$

The converse is also clear. □

By this lemma and Theorem 1.3.11

THEOREM 1.3.34 Let $M = \mathbb{P}^2$ and $B = D_1 \cup \dots \cup D_s$ be a nodal curve, where each D_j $(1 \le j \le s)$ is an irreducible plane curve of degree d_j. Then there is a maximal covering

$$\tilde{\pi} : \widetilde{\mathbb{P}}^2 = \widetilde{\mathbb{P}}^2(D) \longrightarrow \mathbb{P}^2$$

which branches at $D = e_1D_1 + \dots + e_sD_s$ if and only if, for every j $(1 \le j \le s)$, $e_j/(d_j,e_j)$ divides

$$\langle e_1/(d_1,e_1),\dots,\check{e_j/(d_j,e_j)},\dots,e_s/(d_s,e_s)\rangle.$$

In this case, $\tilde{\pi} : \widetilde{\mathbb{P}}^2 \longrightarrow \mathbb{P}^2$ is a finite abelian covering such that

$$G_\pi \simeq \mathbb{Z}\gamma_1 + \dots + \mathbb{Z}\gamma_s,$$

where $d_1\gamma_1 + \dots + d_s\gamma_s = 0$ and $e_j\gamma_j = 0$ for $1 \le j \le s$.

It should be noted that $\widetilde{\mathbb{P}^2}$ is a normal projective surface, by the following theorem of Grauert [20]. For the proof, we also refer Wavrik [63].

THEOREM 1.3.35 (Grauert) Let X and Y be compact complex spaces and $f : X \longrightarrow Y$ be a surjective finite holomorphic map. If Y is projective, then X is also projective.

It should also be noted that Theorem 1.3.34 can be easily generalized to $M = \mathbb{P}^n$ $(n \ge 3)$ and a normally crossing hypersurface B in $\mathbb{P}^n$, by using the following theorem of Zariski [69].

THEOREM 1.3.36 (Zariski) Let B be a hypersurface of $\mathbb{P}^n$ $(n \ge 3)$. Let H be a general hyperplane of $\mathbb{P}^n$. Then

$$\pi_1(\mathbb{P}^n - B, p_o) \simeq \pi_1(H - B\cap H, p_o)$$

Finally we consider abelian branched coverings.

DEFINITION 1.3.37 An abelian branched covering $\pi : X \longrightarrow M$ is called a *maximal abelian covering which branches at* D if (1) π branches at D and (2) $\pi \ge \mu$ for any abelian covering $\mu : Y \longrightarrow M$ which branches at most at D.

If there exists such a covering, it must be unique up to equivalence. We denote it by

$$\tilde{\pi}_{ab} : \tilde{M}_{ab}(D) \longrightarrow M.$$

THEOREM 1.3.38 There is a maximal abelian covering which branches at D if and only if the following condition holds for every j $(1 \le j \le s)$:

$$\text{if } d\gamma_j \in \mathbb{Z}(e_1\gamma_1) + \dots + \mathbb{Z}(e_s\gamma_s), \text{ then } d \equiv 0 \pmod{e_j},$$

where γ_j is regarded as an element of $H_1(M-B;\mathbb{Z})$.

PROOF. Let $\alpha : \pi_1(M-B,p_o) \longrightarrow H_1(M-B;\mathbb{Z})$ be the Hurewicz homomorphism. Put

$$K = \alpha^{-1}(\mathbb{Z}(e_1\gamma_1) + \dots + \mathbb{Z}(e_s\gamma_s)) = J \,.\, [\pi_1(M-B,p_o),\pi_1(M-B,p_o)].$$

Under the notations of Lemma 1.3.5, $K_p = i_*^{-1}K$ contains

$$L_p = [\pi_1(W-B),\ \pi_1(W-B)].$$

Hence

$$\frac{\pi_1(W-B)}{K_p} \simeq \frac{\pi_1(W-B)/L_p}{K_p/L_p} \simeq \frac{H_1(W-B;\mathbb{Z})}{K_p/L_p} .$$

Let $\hat{D}_1 \cup \ldots \cup \hat{D}_t$ be the irreducible decomposition of germ of B at p. Then $H_1(W-B;\ \mathbb{Z})$ is generated by $\hat{\gamma}_1,\ldots,\hat{\gamma}_t$. ($\hat{\gamma}_j$ is a closed path rounding $\hat{D}_j$ once in a counterclockwise direction.) This follows from the fact that the pair (W, W-B) is homeomorphic to the cone $\mathrm{Cone}(\partial\bar{W},\partial\bar{W} \cap B)$ as mentioned before.

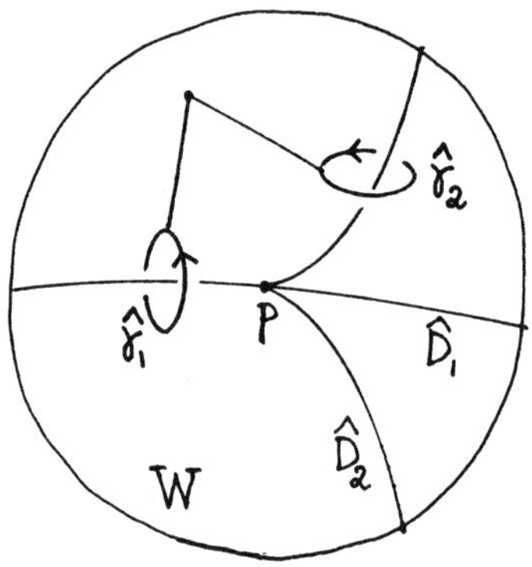

On the other hand, since $i_*(\hat{\gamma}_j)$ $(1 \leq j \leq t)$ is one of $\gamma_1,\ldots,\gamma_s$, $\mathbb{Z}(\hat{e}_1\hat{\gamma}_1)+\ldots+\mathbb{Z}(e_t\hat{\gamma}_t)$ is contained in K_p/L_p. Hence

$$\frac{\pi_1(W-B)}{K_p} \simeq \frac{H_1(W-B;\mathbb{Z})}{K_p/L_p} \simeq \frac{(\mathbb{Z}\hat{\gamma}_1+\ldots+\mathbb{Z}\hat{\gamma}_2)/(\mathbb{Z}(e_1\hat{\gamma}_1)+\ldots+\mathbb{Z}(e_t\hat{\gamma}_t))}{(K_p/L_p)/(\mathbb{Z}(e_1\hat{\gamma}_1)+\ldots+\mathbb{Z}(e_t\hat{\gamma}_t))} .$$

Thus K_p is of finite index in $\pi_1(W-B)$.

By Theorem 1.3.8, there is a covering

$$\pi : X \longrightarrow M$$

which branches at most at D such that

$$K = \pi_*(\pi_1(X-\pi^{-1}(B),q_o)) \qquad (q_o \in \pi^{-1}(p_o)).$$

π is clearly an abelian covering. By the construction, it is also clear that $\pi \geq \mu$ for any abelian covering $\mu : Y \longrightarrow M$ which branches at most at D.

π branches at D if and only if K satisfies Condition 1.3.3, which means that, for every j $(1 \leq j \leq s)$, γ_j has the order e_j as an element of

$$\pi_1(M-B,\ p_o)/K \simeq H_1(M-B;\ \mathbb{Z})/(\mathbb{Z}(e_1\gamma_1)+\ldots+\mathbb{Z}(e_s\gamma_s)). \quad \square$$

EXAMPLES 1.3.39 Put $M = \mathbb{P}^n$ and D_j = an irreducible hypersurface of $\mathbb{P}^n$ of degree d_j $(1 \leq j \leq s)$. Put $B = D_1 \cup \ldots \cup D_s$. As Theorem 1.3.32 (c.f. also Theorem 1.3.36), we have

LEMMA 1.3.40 $H_1(\mathbb{P}^n-B;\ \mathbb{Z})$ is the additive group generated by $\gamma_1,\ldots,\gamma_s$ with the unique relation $d_1\gamma_1+\ldots+d_s\gamma_s = 0$.

Hence $H_1(\mathbb{P}^n-B;\ \mathbb{Z})/(\mathbb{Z}(e_1\gamma_1)+\ldots+\mathbb{Z}(e_s\gamma_s))$ can be identified with the additive group generated by $\gamma_1,\ldots,\gamma_s$ with the relations:

$$d_1\gamma_1+\ldots+d_s\gamma_s = 0, \quad e_1\gamma_1 = \ldots = e_s\gamma_s = 0.$$

(We use γ by abuse of notation.) From this remark, we have as Lemma 1.3.33:

THEOREM 1.3.41 Let D_j be an irreducible hypersurface of $\mathbb{P}^n$ of degree d_j $(1 \leq j \leq s)$. Then there is a maximal abelian covering $\tilde{\pi}_{ab} : \tilde{\mathbb{P}}^n_{ab}(D) \longrightarrow \mathbb{P}^n$ which branches at $D = e_1D_1+\ldots+e_sD_s$ if and only if, for every j $(1 \leq j \leq s)$, $e_j/(d_j,e_j)$ divides

$$\langle e_1/(d_1,e_1),\ldots,\overset{\vee}{e_j/(d_j,e_j)},\ldots,e_s/(d_s,e_s)\rangle.$$

In this case, $\tilde{\pi}_{ab}$ is a finite covering.

EXAMPLE 1.3.42 Let M be a compact Riemann surface of genus g, $B = \{p_1,\ldots,p_c\}$ and $D = e_1(p_1)+\ldots+e_o(p_o)$, where (p_j) is the point divisor of the point $p_j \in M$. Following the notations in section 1.2, $H_1(M-B;\ \mathbb{Z})$ is generated by

$$\alpha_1,\beta_1,\ldots,\alpha_g,\beta_g,\gamma_1,\ldots,\gamma_s$$

with the unique relation

$$\gamma_1+\ldots+\gamma_s = 0.$$

Hence $H_1(M-B;\ \mathbb{Z})/(\mathbb{Z}(e_1\gamma_1)+\ldots+\mathbb{Z}(e_s\gamma_s))$ can be identified with the additive group generated by $\alpha_1,\beta_1,\ldots,\alpha_g,\beta_g,\gamma_1,\ldots,\gamma_s$ with the relations:

$$\gamma_1+\ldots+\gamma_s = 0, \quad e_1\gamma_1 = \ldots = e_s\gamma_s = 0.$$

(We use α,β,γ by abuse of notation.) Hence, from Lemma 1.3.33,

THEOREM 1.3.43 Let M be a compact Riemann surface of genus g. Then there is a maximal abelian covering $\tilde{\pi}_{ab} : \tilde{M}_{ab}(D) \longrightarrow M$ which branches at $D = e_1(p_1) + \ldots + e_s(p_s)$ if and only if, for every j $(1 \leq j \leq s)$, e_j divides

$$\langle e_1, \ldots, \check{e}_j, \ldots, e_s \rangle.$$

Note that $\tilde{\pi}_{ab}$ in this theorem is *not* a finite covering if $g \geq 1$.

As for finite abelian coverings, we have:

THEOREM 1.3.44 There is a one-to-one correspondence $\pi \longmapsto K = K(\pi)$ between isomorphism classes of finite abelian coverings $\pi : X \longrightarrow M$ of a complex manifold M which branches at most at D and subgroups K of finite index of

$$G_{ab} = H_1(M-B; \mathbb{Z})/(\mathbb{Z}(e_1\gamma_1) + \ldots + \mathbb{Z}(e_s\gamma_s)).$$

The correspondence satisfies (1) $G_\pi \simeq G_{ab}/K(\pi)$, (2) $\pi_1 \leq \pi_2$ if and only if $K(\pi_1) \supset K(\pi_2)$ and (3) π branches at D if and only if $K(\pi)$ satisfies the following condition: for every j $(1 \leq j \leq s)$,

$$\text{if } d\gamma_j \in K(\pi), \quad \text{then} \quad d \equiv 0 \pmod{e_j}.$$

PROOF. This follows from Theorem 1.3.9 and the isomorphism:

$$\pi_1(M-B, p_o)/J \cdot [\pi_1(M-B, p_o), \pi_1(M-B, p_o)]$$
$$\simeq H_1(M-B; \mathbb{Z})/(\mathbb{Z}(e_1\gamma_1) + \ldots + \mathbb{Z}(e_s\gamma_s)). \quad \square$$

1.4 Examples of finite branched coverings

In this section, we give some concrete examples of finite branched coverings.

1°. The first example is the typical cyclic covering. Let M be an n-dimensional compact complex manifold and

$$\hat{\pi} : F \longrightarrow M$$

be a holomorphic line bundle on M. Take an open covering $\{U_\alpha\}$ of M such that F is locally trivial on U_α. Let $f_{\alpha\beta}$ be the transition functions of F with respect to the local trivialization. We write $F = \{f_{\alpha\beta}\}$.

For a positive integer e greater than 1, let $\xi = \{\xi_\alpha\}$ be a holomorphic section of $F^{\otimes e}$ (e-times tensor product of F) and let (ξ) be the zero divisor of ξ. Suppose that (ξ) has no multiple component and let

$$(\xi) = D_1 + \ldots + D_s$$

be the decomposition of (ξ) into prime divisors. Put

$$B = D_1 \cup \dots \cup D_s$$

and

$$D = eD_1 + \dots + eD_s.$$

The bundle space F can be written as

$$F = (\cup_\alpha(U_\alpha \times \mathbb{C}))/\sim,$$

where $(p_\alpha, z_\alpha) \sim (p_\beta, z_\beta)$ if $p_\alpha = p_\beta \in U_\alpha \cap U_\beta$ and $z_\alpha = f_{\alpha\beta}(p_\beta)z_\beta$.

Now we put

$$X_\alpha = \{(p_\alpha, z_\alpha) \in U_\alpha \times \mathbb{C} \mid z_\alpha^e = \xi_\alpha(p_\alpha)\}.$$

We regard X_α as a hypersurface of $\hat{\pi}^{-1}(U_\alpha)$. Then we have

$$X_\alpha = X_\beta \quad \text{in} \quad \hat{\pi}^{-1}(U_\alpha \cap U_\beta).$$

Thus X_α's can be patched up and define a hypersurface X of the bundle space F.

$$\pi = \hat{\pi}|X : X \longrightarrow M$$

is then a surjective proper finite holomorphic map. We have

LEMMA 1.4.1 (1) $\pi : X - \pi^{-1}(B) \longrightarrow M-B$ is unramified, (2) for any point $p \in B$, the fibre $\pi^{-1}(p)$ consists of one point, (3) X is singular at (p_α, z_β) if and only if $p_\alpha \in \mathrm{Sing}B$ and $z_\alpha = 0$, and (4) X is an irreducible normal complex space.

PROOF. (1) - (3) are trivial. Since X is a hypersurface of a complex manifold and codim $\mathrm{Sing}X \geq 2$ (by (3)), X is a normal complex space. In order to show that X is irreducible, it suffices to show that X-SingX is connected. It suffices then to show that any points q_1 and q_2 in $X - \pi^{-1}(B)$ can be connected by a path in X-SingX. Put $p_1 = \pi(q_1)$ and $p_2 = \pi(q_2)$. Take a point $p_3 \in B-\mathrm{Sing}B$. Let β_1 (resp. β_2) be a path in M-SingB from p_1 to p_3 (resp. from p_2 to p_3) which is contained in M-B except the end point p_3. Then there is a path $\hat{\beta}_1$ (resp. $\hat{\beta}_2$) in $X-\pi^{-1}(\mathrm{Sing}B)$ from q_1 to $q_3 = \pi^{-1}(p_3)$, (resp. from q_2 to q_3) such that $\pi(\hat{\beta}_1) = \hat{\beta}_1$ (resp. $\pi(\hat{\beta}_2) = \beta_2$). Now $\hat{\beta}_1\hat{\beta}_2^{-1}$ is a path from q_1 to q_2. This proves (4). □

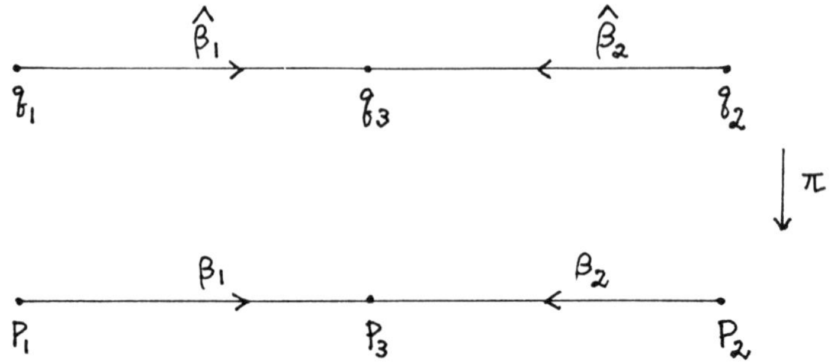

The covering transformation group G of $\pi : X \longrightarrow M$ is isomorphic to the cyclic group $(\mathbb{Z}/e\mathbb{Z})$. Thus $\pi : X \longrightarrow M$ is a cyclic covering of degree e which branches at D.

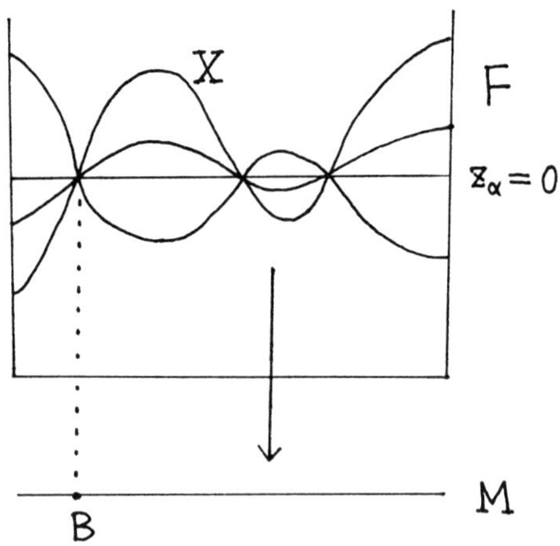

If $M = \mathbb{P}^n$ and F = the hyperplane bundle, then ξ can be identified with a homogeneous polynomial $G = G(Z_0,\ldots,Z_n)$ of degree e. In this case, X can be identified with the hypersurface

$$X = \{(Z_0:\ldots:Z_n:Z_{n+1}) \in \mathbb{P}^{n+1} \mid Z_{n+1}^e - G(Z_0,\ldots,Z_n) = 0\}$$

of $\mathbb{P}^{n+1}$.

2°. Let $\hat{\pi} : F \longrightarrow M$ be as in 1°. Let $\xi_1 = \{\xi_{1\alpha}\}$, $\xi_2 = \{\xi_{2\alpha}\},\ldots,\xi_e = \{\xi_{e\alpha}\}$ be holomorphic sections of F, $F^{\otimes 2},\ldots,F^{\otimes e}$, respectively. Put

$$X_\alpha = \{(p_\alpha, z_\alpha) \in U_\alpha \times \mathbb{C} \mid z_\alpha^e + \xi_{1\alpha}(p_\alpha)z_\alpha^{e-1} + \xi_{2\alpha}(p_\alpha)z_\alpha^{e-2} + \ldots + \xi_{e\alpha}(p_\alpha) = 0\}$$

and regard it as a hypersurface of $\hat{\pi}^{-1}(U_\alpha)$. Then X_α's can be patched up and define a hypersurface X of F.

$$\pi = \hat{\pi}|X : X \longrightarrow M$$

is a surjective proper finite holomorphic map. (Conversely, it can be shown that, if X is a hypersurface of F such that $\pi = \pi|X : X \longrightarrow M$ is a surjcctive proper finite map, then X is given as above, (see Wavrik [63]).)

Suppose that X is irreducible. Then $\pi : X \longrightarrow M$ is a (not necessarily Galois) finite branched covering of degree e, whose branch locus is the discriminant locus

$$B = \{p \in M \mid D(\xi_1(p),\ldots,\xi_e(p)) = 0\}$$

of the defining equation of X.

3°. Let M be an n-dimensional compact complex manifold and $F_1,\ldots,F_s$ be holomorphic line bundles on M. Let e_j $(1 \leq j \leq s)$ be positive integers greater than 1 and ξ_j $(1 \leq j \leq s)$ be holomorphic sections of $F_j^{\otimes e_j}$, $(1 \leq j \leq s)$, respectively. Suppose that the zero divisors $D_j = (\xi_j)$ of ξ_j $(1 \leq j \leq s)$ do not have multiple components and any two of them do not have a common component. Put

$$B = D_1 \cup \ldots \cup D_s,$$

and

$$D = e_1D_1 + \ldots + e_sD_s.$$

Let $\pi_j : X_j \longrightarrow M$ be the cyclic covering defined in 1° with respect to the data (F_j,e_j,ξ_j). Consider the fibre product

$$\pi' : X' = X_1 \times_M \ldots \times_M X_s \longrightarrow M$$

of $X_1,\ldots,X_s$. Then X' is a closed complex subspace of

$$F = F_1 \oplus \ldots \oplus F_s.$$

π' is a surjective proper finite map. Note that $\pi'^{-1}(M\text{-Sing}B)$ is non-singular.

<u>LEMMA 1.4.2</u> X' is irreducible.

<u>PROOF</u>. It suffices to show that $\pi'^{-1}(M\text{-Sing}B)$ is connected. Hence it suffices to show that any two points $(\hat{p}_1,\ldots,\hat{p}_s)$ and $(\hat{q}_1,\ldots,\hat{q}_s)$ in $\pi'^{-1}(M\text{-}B)$ can be connected by a path in $\pi'^{-1}(M\text{-Sing}B)$. Put

$$p = \pi'(\hat{p}_1,\dots,\hat{p}_s) \quad \text{and} \quad q = \pi'(\hat{q}_1,\dots,\hat{q}_s).$$

Take points $r_j \in D_j$-SingB and $\hat{r}_j \in \pi_j^{-1}(r_j)$ for $1 \leq j \leq s$. Note that $\hat{r}_j$ is uniquely determined by r_j.

Let β_j be a path in M-SingB which connects from r_{j-1} to r_j for $1 \leq j \leq s+1$. (For convenience, we put $p = r_o$ and $q = r_{s+1}$.)

Let $\hat{\delta}_j$ be a (uniquely determined) path in π_j^{-1}(M-SingB) which starts from p_j such that

$$\pi_j(\hat{\delta}_j) = \beta_1\dots\beta_j, \quad \text{for } 1 \leq j \leq s.$$

Let $\hat{\varepsilon}_j$ be a uniquely determined) path in π_j^{-1}(M-SingB) which starts from q_j such that

$$\pi_j(\hat{\varepsilon}_j) = \beta_{s+1}^{-1}\dots\beta_{j+1}^{-1} \quad \text{for} \quad 1 \leq j \leq s.$$

Then $\hat{\delta}_j$ and $\hat{\varepsilon}_j$ have the same end point $\hat{r}_j$.

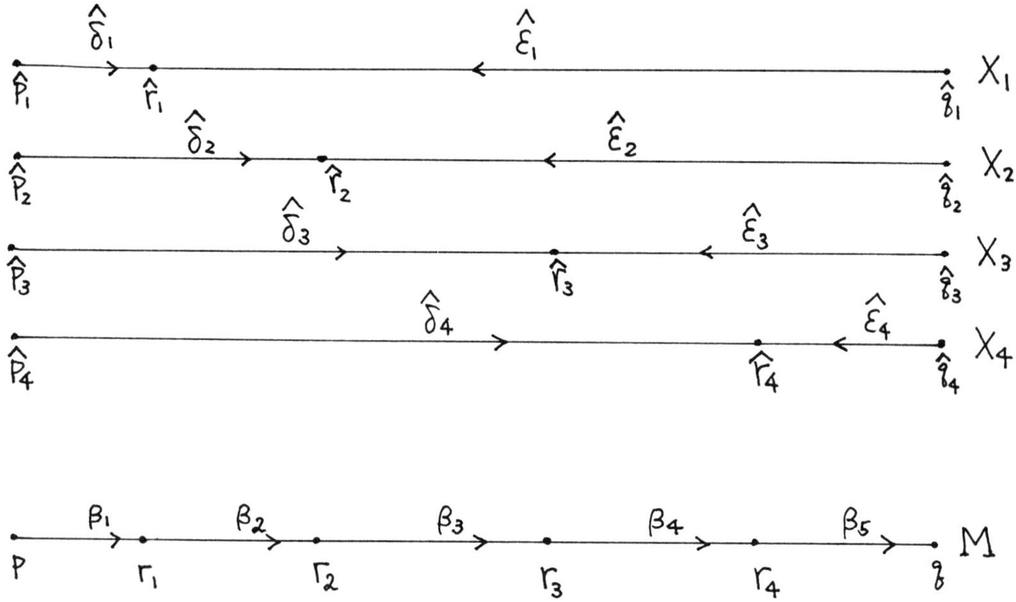

Thus $\hat{\alpha}_j = \hat{\delta}_j\hat{\varepsilon}_j^{-1}$ is a path in π_j^{-1}(M-SingB) which connects from $\hat{p}_j$ to $\hat{q}_j$ for $1 \leq j \leq s$. Now

$$\hat{\alpha} = (\hat{\alpha}_1,\dots,\hat{\alpha}_s)$$

is a path in π'^{-1}(M-SingB) which connects from $(\hat{p}_1,\dots,\hat{p}_s)$ to $(\hat{q}_1,\dots,\hat{q}_s)$. □

Let $\rho : X \longrightarrow X'$ be the normalization of X'. Put $\pi'\rho = \pi$. Then

$$\pi : X \longrightarrow M$$

is a finite abelian covering of M which branches at D. The covering transformation group G_π of π is isomorphic to

$(\mathbb{Z}/e_1\mathbb{Z})\oplus\ldots\oplus(\mathbb{Z}/e_s\mathbb{Z})$.

<u>LEMMA 1.4.3</u> For a point $p \in B$, suppose that (1) every D_j passing through p is non-singular at p and (2) B is normally crossing at p. Then X' is non-singular at every point of $\pi'^{-1}(p)$.

<u>PROOF</u>. Let W be an open neighbourhood of p with coordinate system $(w_1,\ldots,w_n)$ such that $p = (0,\ldots,0)$,

$$B \cap W = \{(w_1,\ldots,w_n) \in W \mid w_1\ldots w_m = 0\} \qquad (m \leq n)$$

and such that every F_j is trivial on W. We may suppose that

$$D_j \cap W = \{(w_1,\ldots,w_n) \in W \mid w_j = 0\}, \qquad \text{for } 1 \leq j \leq m.$$

$\pi'^{-1}(W)$ is then given by the equations

$$z_1^{e_1} - w_1 = 0,\ldots,z_m^{e_m} - w_m = 0, \quad z_{m+1}^{e_{m+1}} - \xi_{m+1}(w) = 0,\ldots,z_s^{e_s} - \xi_s(w) = 0$$

in $W \times \mathbb{C}^s$. Note that $\xi_{m+1}(0) \neq 0,\ldots,\xi_s(0) \neq 0$. Take $u^0_{m+1},\ldots,u^0_s$ such that $(u^0_j)^{e_j} = \xi_j(0)$ for $m+1 \leq j \leq s$, and put $u^0 = (0,\ldots,0,u^0_{m+1},\ldots,u^0_s) \in \mathbb{C}^s$. Consider the holomorphic map

$$\alpha : (w_1,\ldots,w_n,z_1,\ldots,z_s) \in W \times \mathbb{C}^s$$

$$\longmapsto (z_1^{e_1} - w_1,\ldots,z_m^{e_m}-w_m,\ z_{m+1}^{e_{m+1}}-\xi_{m+1}(w),\ldots,z_s^{e_s}-\xi_s(w)) \in \mathbb{C}^s$$

and its Jacobian matrix

$$J_\alpha(0,u^0) = \left(\begin{array}{c:c:c:c} \overset{(m)}{\begin{matrix}-1 & & \\ & \ddots & \\ & & -1\end{matrix}} & \overset{(n-m)}{0} & \overset{(m)}{0} & \overset{(s-m)}{0} \\ \hdashline * & * & * & \begin{matrix} e_{m+1}(u^0_{m+1})^{e_{m+1}-1} & & \\ & \ddots & \\ & & e_s(u^0_s)^{e_s-1}\end{matrix} \end{array}\right) \begin{matrix}(m)\\ \\ (s-m)\end{matrix}$$

at $(0,u^0)$. This matrix has the rank s. Hence X' is non-singular at $(0,u^0)$. □

By Lemmas 1.4.2 and 1.4.3, we have

PROPOSITION 1.4.4 If (1) every D_j is non-singular and irreducible and (2) B is normally crossing, then X' is a compact complex manifold. Hence X' = X and π' = π. Thus $\pi : X \longrightarrow M$ is an abelian covering of normally crossing type.

Under the conditions of the proposition, the canonical bundle K_X of X = X' can be computed by the adjunction formula:

$$K_X = \pi^* K_M + [C_\pi],$$

where

$$C_\pi = (e_1 - 1)\hat{D}_1 + \ldots + (e_s - 1)\hat{D}_s$$

$$\hat{D}_j = \pi^{-1}(D_j) \quad \text{for} \quad 1 \leq j \leq s.$$

Here, we use addition, +, instead of tensor product, ⊗.
($\hat{D}_j$ is not $\pi^*(D_j)$, but the set theoretic pull back of D_j.)

In particular, if

$$F_1 = \ldots = F_s = F,$$

then, as is easily seen,

$$[\hat{D}_j] = \pi^* F \quad \text{for} \quad 1 \leq j \leq s.$$

Hence

$$K_X = \pi^*(K_M + (e_1 + \ldots + e_s - s)F).$$

In particular, if $M = \mathbb{P}^n$ and F is the hyperplane bundle, then

$$K_X = (e_1 + \ldots + e_s - s - n - 1)\pi^* F.$$

Hence K_X is ample if and only if

$$e_1 + \ldots + e_s \geq s + n + 2.$$

Note that, if $M = \mathbb{P}^2$ and F is the hyperplane bundle, then our X is nothing but $\tilde{\mathbb{P}}^2(D)$ for the case $d_1 = e_1, \ldots, d_s = e_s$, in 3° of section 1.3. In fact, they have the covering transformation groups over $\mathbb{P}^2$ both isomorphic to $(\mathbb{Z}/e_1\mathbb{Z}) \oplus \ldots \oplus (\mathbb{Z}/e_s\mathbb{Z})$. In particular, X is simply connected by Theorem 1.3.10. Moreover, following Hirzebruch [30], we can compute $c_1(X)^2 = K_X^2$ and $c_2(X) = E(X)$:

$$c_1(X)^2 = K_X^2 = (e_1 + \ldots + e_s - s - 3)^2 a_s,$$

and

$$c_2(X) = E(X) = \{(a_1-s)(a_1-3) + 3-a_2 + s(s-1)/2\}a_s,$$

where

$$a_1 = e_1+\ldots+e_s,$$

$$a_2 = \Sigma_{j<k} e_j e_k,$$

$$\vdots$$

$$a_s = e_1\ldots e_s.$$

If (M,F) is a principally polarized abelian variety, that is, if F is a positive line bundle on a complex n-torus M such that the Chern class c(F) has the elementary divisors (1,...,1), (see Griffiths and Harris [22]), then

$$K_X = (e_1+\ldots+e_s - s)\pi^*F.$$

Hence, K_X is always ample in this case.

In particular, if n = 2, then

$$c_1(X)^2 = K_X^2 = 2(a_1-s)^2 a_s,$$

and

$$c_2(X) = E(X) = \{2a_1^2-2a_2-2a_1 s+s(s-1)\}a_s.$$

4°. Let M be an n-dimensional projective manifold and f be a meromorphic (i.e. rational) function on M. We may write

$$f = \xi_2/\xi_1,$$

where ξ_1 and ξ_2 are holomorphic sections of a line bundle F on M such that the supports of the divisors $D_1 = (\xi_1)$ and $D_2 = (\xi_2)$ have no common component. (D_1 and D_2 are the polar and zero divisors of f, respectively.) For a positive integer e greater than 1, consider the hypersurface

$$X' = \{(p,z) \in M \times \mathbb{P}^1 \mid z^e = f(p)\}$$

$$= \{(p,(Z_1:Z_2)) \in M \times \mathbb{P}^1 \mid Z_2^e\xi_1(p) - Z_1^e\xi_2(p) = 0\}$$

of $M \times \mathbb{P}^1$. The projection

$$\pi' : X' \longrightarrow M,$$

induced from the projection $M \times \mathbb{P}^1 \longrightarrow M$, is a surjective proper holomorphic map. π' is not a finite map in general. In fact

$$\pi'^{-1}(p) \simeq \mathbb{P}^1 \quad \text{(biholomorphic) for } p \in \mathrm{Supp}(D_1) \cap \mathrm{Supp}(D_2)$$

and

$$\pi'^{-1}(p) \text{ is a finite set for } p \in M\text{-}\mathrm{Supp}(D_1) \cap \mathrm{Supp}(D_2).$$

Moreover, X' may not be irreducible in general. For example, put

$$M = \mathbb{P}^1, \quad f(x) = x^2 \quad \text{and} \quad e = 2.$$

Then

$$X' = \{(x,z) \in \mathbb{P}^1 \times \mathbb{P}^1 \mid z^2 = x^2\}$$

is not irreducible. But

LEMMA 1.4.5 If either of D_1 and D_2 has no multiple component, then X' is irreducible.

This lemma can be proved in a similar way to Lemma 1.4.1, by taking

$$q_1, q_2 \in X' - \pi'^{-1}(\mathrm{Supp}D_1 \cap \mathrm{Supp}D_2),$$

$$p_3 \in \mathrm{Supp}D_1 - \mathrm{Sing}D_1 - \mathrm{Supp}D_1 \cap \mathrm{Supp}D_2.$$

For simplicity, suppose that both D_1 and D_2 have no multiple component. Let $\mathbb{C}(X')$ (resp. $\mathbb{C}(M)$) be the field of all meromorphic (i.e. rational) functions of X' (resp. of M). Then $\mathbb{C}(X')$ is a cyclic extension of $\mathbb{C}(M)$. In fact

$$\mathbb{C}(X') = \mathbb{C}(M)(f^{1/e}).$$

Let

$$\pi : X \longrightarrow M$$

be the $\mathbb{C}(X')$-*normalization of* M (see Iitaka [33]). Then this is a cyclic covering of M which branches at

$$D = eD_1 + eD_2$$

such that $G_\pi \simeq \mathbb{Z}/e\mathbb{Z}$.

For example, if $M = \mathbb{P}^2$ with a homogeneous coordinate system $(Y_0:Y_1:Y_2)$ and $f = Y_2/Y_1$, $(\xi_1 = Y_1,\ \xi_2 = Y_2)$, then

$$X' = \{((Y_0:Y_1:Y_2), (Z_1:Z_2)) \in \mathbb{P}^2 \times \mathbb{P}^1 \mid Z_2^e Y_1 - Z_1^e Y_2 = 0\}$$

is a non-singular surface, called the *Hirzebruch surface*. X can be obtained from X' by blowing down $(1:0:0) \times \mathbb{P}^1$ to a point, which is a unique singular point.

5°. Let $f_1,\ldots,f_e$ $(e \geq 2)$ be meromorphic functions on a projective manifold M. We may write

$$f_j = \xi_j/\xi_0 \quad \text{for} \quad 1 \leq j \leq e,$$

where $\xi_0,\ldots,\xi_e$ are holomorphic sections of a line bundle F on M such that the supports of divisors

$$D_0 = (\xi_0),\ldots,D_e = (\xi_e)$$

have no common component. Consider the hypersurface

$$X' = \{(p,z) \in M \times \mathbb{P}^1 \mid z^e+f_1(p)z^{e-1} + \ldots + f_e(p) = 0\}$$
$$= \{(p,(Z_1:Z_2)) \in M \times \mathbb{P}^1 \mid \xi_0(p)Z_2^e+\xi_1(p)Z_2^{e-1}Z_1+\ldots+\xi_e(p)Z_1^e = 0\}$$

of $M \times \mathbb{P}^1$. Suppose that X' is irreducible. The projection

$$\pi' = X' \longrightarrow M,$$

induced from the natural projection $M \times \mathbb{P}^1 \longrightarrow M$, is a surjective proper holomorphic map. The natural projection $(p,z) \in M \times \mathbb{P}^1 \longmapsto z \in \mathbb{P}^1$ induces a meromorphic function z on X'. z can be regarded as a multi-valued meromorphic function on M. If $M = \mathbb{P}^n$, then this multi-valued function is classically called an algebraic function. In this book, however, by an *algebraic function* is simply meant a meromorphic (i.e. rational) function on an irreducible normal (quasi-)projective variety.

Let

$$\pi : X \longrightarrow M$$

be the $\mathbb{C}(X')$-normalization of M. Then this is a finite branched covering of degree e, whose branch locus is the discriminant locus

$$B = \{p \in M \mid D(\xi_0(p),\ldots,\xi_e(p)) = 0\}$$

of the defining equation of X'.

6°. Let F be a line bundle on an n-dimensional projective manifold M and $\xi_1,\ldots,\xi_s$ be holomorphic sections of F on M such that $D_1 = (\xi_1),\ldots,D_s = (\xi_s)$ have no multiple component and any two of them have no common component. For a positive integer e $(e \geq 2)$, put

$$B = D_1 \cup \ldots \cup D_s,$$

$$D = eD_1 + \ldots + eD_s.$$

Consider the Kummer extension

$$K = \mathbb{C}(M)((\xi_1/\xi_s)^{1/e},\ldots,(\xi_{s-1}/\xi_s)^{1/e})$$

of the field $\mathbb{C}(M)$. Let

$$\pi : X \longrightarrow M$$

be the K-normalization of M. Then π is a finite abelian covering which branches at D.

For simplicity, *suppose that*

$$D_1 \cap \ldots \cap D_s = \emptyset.$$

Consider the algebraic set

$$X' = \{(p,(Z_1:\ldots:Z_s)) \in M \times \mathbb{P}^{s-1} \mid Z_j^e\xi_k(p)-Z_k^e\xi_j(p) = 0,\ 1 \leq j,\ k \leq s\}$$

and the projection

$$\pi' : X' \longrightarrow M$$

induced from the natural projection $M \times \mathbb{P}^{s-1} \longrightarrow M$. X' is in fact the fibre product $X' = M \times_{\mathbb{P}^{s-1}} \mathbb{P}^{s-1}$:

$$\begin{array}{ccc} X' & \longrightarrow & \mathbb{P}^{s-1} \\ {\scriptstyle \pi'}\downarrow & & \downarrow{\scriptstyle A} \\ M & \xrightarrow{\ \Phi\ } & \mathbb{P}^{s-1} \end{array}$$

where

$$\Phi : p \in M \longmapsto (\xi_1(p):\ldots:\xi_s(p)) \in \mathbb{P}^{s-1}$$

and

$$A : (Y_1:\ldots:Y_s) \in \mathbb{P}^{s-1} \longmapsto (Y_1^e:\ldots:Y_s^e) \in \mathbb{P}^{s-1}.$$

Then π' is clearly a surjective finite map.

<u>LEMMA 1.4.6</u> Under the assumption $D_1 \cap \ldots \cap D_s = \emptyset$, (1) X' is irreducible, (2) $\pi : X \longrightarrow M$ is equivalent to $\pi'\rho : X'_n \longrightarrow M$, where $\rho : X'_n \longrightarrow M$, is the normalization of X' and (3) $G_\pi \simeq (\mathbb{Z}/e\mathbb{Z})^{s-1}$.

PROOF (1) can be proved in a similar way to Lemma 1.4.2. (2) is trivial. (3) follows from (2). □

A similar proof to that of Lemma 1.4.3 shows

LEMMA 1.4.7 Assume that $D_1 \cap \ldots \cap D_s = \emptyset$. For a point $p \in B$, suppose that (1) every D_j passing through p is non-singular at p and (2) B is normally crossing at p. Then X' is non-singular at every point of $\pi'^{-1}(p)$.

By this lemma,

PROPOSITION 1.4.8 Assume that $D_1 \cap \ldots \cap D_s = \emptyset$. If (1) every D_j is non-singular and irreducible and (2) B is normally crossing, then X' is a projective manifold. Hence $X' = X$ and $\pi' = \pi$. Thus $\pi : X \longrightarrow M$ is an abelian covering of simple normally crossing type.

For the rest of this section, we discuss under the following assumption:

ASSUMPTION (1) $D_1 \cap \ldots \cap D_s = \emptyset$, (2) $\dim M = 2$, (3) every D_j is a non-singular irreducible curve on M and (4) $B = D_1 \cup \ldots \cup D_s$ has only ordinary multiple points as its singularity.

This is the case that Hirzebruch [30] treated recently.

For a point $p \in B$, we denote by $m_p = m_p(B)$ the multiplicity of B at p. Put

$$\Delta : \{p \in B \mid m_p \geq 3\}.$$

We blow up every point $p \in \Delta$ and get a projective surface $\hat{M}$ and a proper bimeromorphic holomorphic map

$$\tau : \hat{M} \longrightarrow M.$$

For $p \in \Delta$, $E_p = \tau^{-1}(p)$ is an exceptional curve of the first kind, that is, $E_p \simeq \mathbb{P}^1$ and $E_p^2 = -1$. Note that the curve

$$\hat{B} = (\bigcup_{j=1}^{s} \hat{D}_j) \cup (\bigcup_{P\in\Delta} E_p)$$

has only nodes as its singularity, where $\hat{D}_j$ is the strict transform of D_j.

For a point $p \in \Delta$, we may assume that $p \in D_1 \cap \ldots \cap D_m$, say, where $m = m_p$ (≥ 3). We can take a local coordinate system (w_1, w_2) on an open neighbourhood W of p such that

$$p = (0,0),$$

$$D_1 \cap W = \{(w_1,w_2) \in W \mid w_1 = 0\},$$

$$D_2 \cap W = \{(w_1,w_2) \in W \mid w_2 = 0\},$$

$$D_3 \cap W = \{(w_1,w_2) \in W \mid w_1 - \rho_3 w_2 + g_3(w_1,w_2) = 0\},$$

$$\vdots$$

$$D_m \cap W = \{(w_1,w_2) \in W \mid w_1 - \rho_m w_2 + g_m(w_1,w_2) = 0\},$$

where $\rho_3,\ldots,\rho_m$ are mutually distinct non-zero complex numbers, and the degrees of $g_j(w_1,w_2)$ in the power series expansion with respect to (w_1,w_2) is greater than 1.

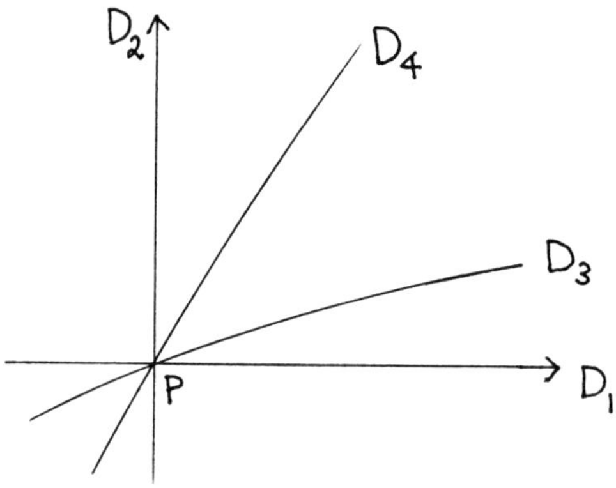

Then we may assume that $\pi'^{-1}(W)$ $(\subset X')$ is defined by the following equations:

$$z_1^e - w_1 = 0,$$

$$z_2^e - w_2 = 0,$$

$$z_j^e - w_1 + \rho_j w_2 - g_j(w_1,w_2) = 0 \quad \text{for} \quad 3 \leq j \leq m,$$

$$z_k^e - f_k(w_1,w_2) = 0 \quad \text{for} \quad m+1 \leq k \leq s-1$$

where

$$f_k(0,0) \neq 0 \quad \text{for} \quad m+1 \leq k \leq s-1.$$

Consider the point $q = (z_1,\ldots,z_{s-1}) = (0,\ldots,0,z^o_{m+1},\ldots,z^o_{s-1}) \in \pi'^{-1}(p)$, where $(z^o_k)^e = f_k(0,0)$ for $m+1 \leq k \leq s-1$. Eliminating w_1, w_2 and z_k

$(m+1 \leq k \leq s-1)$ from the above equations, we get the defining equations of X' around q as follows:

$$z_j^e - z_1^e + \rho_j z_2^e - g_j(z_1^e, z_2^e) = 0 \quad \text{for} \quad 3 \leq j \leq m.$$

Thus X' around q is an intersection of hypersurfaces of Fermat type.

We blow up the $(z_1, \ldots, z_{s-1})$-space at q; we put

$$z_2 = z_1 v_2, \ldots, z_m = z_1 v_m.$$

Then the strict transform $\hat{X}$ of X' can be given by the equations

$$v_j^e - 1 + \rho_j v_2^e - z_1^e \hat{g}_j(z_1, v_2^e) = 0 \quad \text{for} \quad 3 \leq j \leq m, \qquad \ldots(*)$$

where $\hat{g}_j$ are holomorphic functions.

The Jacobian matrix at $(z_1, v_2, \ldots, v_m) = (0, v_2, \ldots, v_m) \in \hat{X}$ of the map

$$(z_1, v_2, \ldots, v_m) \longmapsto (v_3^e - 1 + \rho_3 v_2^e - z_1^e \hat{g}_3(z_1, v_2^e), \ldots,$$
$$v_m^e - 1 + \rho_m v_2^e - z_1^e \hat{g}_m(z_1, v_2^e)) \in \mathbb{C}^{m-2}$$

is given by

$$J = \begin{pmatrix} 0 & \rho_3 e v_2^{e-1} & e v_3^{e-1} & & 0 \\ \cdot & \cdot & & \cdot & \\ \cdot & & & & \cdot \\ & \cdot & & & \\ \cdot & \cdot & & & \cdot \\ 0 & \rho_m e v_2^{e-1} & 0 & & e v_m^{e-1} \end{pmatrix}$$

If $v_2 = 0$, then $v_j^e = 1$ for $3 \leq j \leq m$ by $(*)$. Hence J has the rank $m-2$. If $v_3 = 0$, then $v_2^e = 1/\rho_3$ by $(*)$. In this case, $v_4 \neq 0, \ldots, v_m \neq 0$. In fact, if $v_4 = 0$, say, then $v_2^e = 1/\rho_4$ by $(*)$ again. Hence $\rho_3 = \rho_4$, a contradiction. Hence, J has the rank $m-2$ again. In a similar way, if $v_j = 0$ for some j with $4 \leq j \leq m$, then J has the rank $m-2$. If $v_3 \neq 0, \ldots, v_m \neq 0$, then J clearly has the rank $m-2$. Hence, J has the maximal rank in every case.

A similar assertion holds if we put

$$z_1 = z_2 v_1, \quad z_3 = z_2 v_3, \ldots, z_m = z_2 v_m.$$

Thus the strict transform $\hat{X}$ of X' is a non-singular projective surface. Let $\alpha : \hat{X} \longrightarrow X'$ be the natural projection. The map

$$(z_1, v_2, \ldots, v_m) \longmapsto (w_1, v_2) = (z_1^e, v_2)$$

induces a surjective proper finite map

$$\hat{\pi} : \hat{X} \longrightarrow \hat{M}$$

such that the diagram

$$\begin{array}{ccc} \hat{X} & \xrightarrow{\alpha} & X' \\ \hat{\pi}\downarrow & & \downarrow \pi' \\ \hat{M} & \xrightarrow{\tau} & M \end{array}$$

commutes.

PROPOSITION 1.4.9 $\hat{\pi} : \hat{X} \longrightarrow \hat{M}$ is a finite abelian covering of simple normally crossing type which branches at

$$\hat{D} = e\hat{D}_1 + \dots + e\hat{D}_s + \sum_{p \in \Delta} eE_p,$$

such that the covering transformation group $G_{\hat{\pi}}$ is isomorphic to $(\mathbb{Z}/e\mathbb{Z})^{s-1}$.

PROOF. Take a point $(z_1, v_2, \dots, v_m) = (0, v_2^o, \dots, v_m^o) \in \hat{\pi}^{-1}(E_p)$ such that $v_2^o \neq 0$ and $(v_2^o)^e \neq 1/\rho_j$ for $3 \leq j \leq m$. Then, around this point, (z_1, v_2) can be taken as a local coordinate system in $\hat{X}$ and $\hat{\pi}$ can be given by

$$\hat{\pi} : (z_1, v_2) \longmapsto (w_1, v_2) = (z_1^e, v_2).$$

Hence the ramification index at E_p is e.

$G_{\hat{\pi}}$ is isomorphic to $(\mathbb{Z}/e\mathbb{Z})^{s-1}$, since the above diagram commutes. □

Following Hirzebruch [30], we can compute the numerical invariants c_1^2 and c_2 of the projective surface $\hat{X}$. Since

$$K_{\hat{M}} = \tau^* K_M + \sum_{p \in \Delta} [E_p],$$

we have

$$eK_{\hat{X}} = \hat{\pi}^*(e\tau^* K_M + \sum_{p \in \Delta} (2e-1)[E_p] + \sum_{j=1}^{s} (e-1)[\hat{D}_j]).$$

Hence

$$c_1(\hat{X})^2/e^{s-3} = (eK_M + (e-1)B)^2 - \sum_{m \geq 3} (2e-1-m(e-1))^2 t_m,$$

where t_m is the number of m-ple points on B. Also,

$$c_2(\hat{X})/e^{s-3} = e^2(c_2(M) - \sum_{j=1}^{s} (2-2g_j) + f_1 - f_0) +$$

$$e(\sum_j (2-2g_j) + 2f_0 - 2f_1) + (f_1 - t_2),$$

where g_j is the genus of D_j, $f_0 = \sum_{m\geq 2} t_m$ and $f_1 = \sum_{m\geq 2} mt_m$.

In particular, if $M = \mathbb{P}^2$ and every D_j is a line on $\mathbb{P}^2$, then

$$c_1(\hat{X})^2/e^{s-3} = ((e-1)s-3e)^2 - \sum_{m\geq 3} (2e-1-m(e-1))^2 t_m,$$

and

$$c_2(\hat{X})/e^{s-3} = e^2(3-2s+f_1-f_0) + e(2s+2f_0-2f_1) + (f_1-t_2).$$

Now we recall the celebrated

THEOREM 1.4.10 (Miyaoka [42] and Yau [68]). For a complex surface S of general type,

$$c_1(S)^2 \leq 3c_2(S).$$

Moreover, the equality holds if and only if the universal covering space of S is biholomorphic to the unit ball

$$\mathbb{B}^2 = \{(z_1,z_2) \in \mathbb{C}^2 \mid |z_1|^2 + |z_2|^2 < 1\}.$$

Hirzebruch [30] gave several examples of $\hat{X}$ such that the equality $c_1(\hat{X})^2 = 3c_2(\hat{X})$ holds. Among them, the following is the simplest.

Let $B = D_1 \cup \ldots \cup D_6$ ($s = 6$) be the configuration of 6 lines on $\mathbb{P}^2$ as in the following picture:

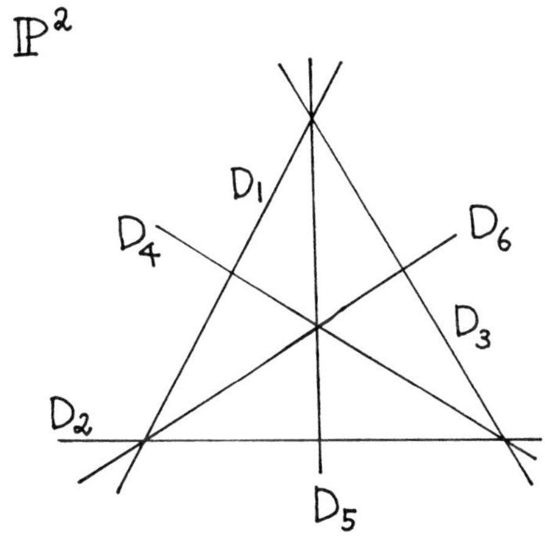

In this case

$$c_1(\hat{X})^2/e^3 = 5(e-2)^2,$$

$$3c_2(\hat{X})/e^3 = 6e^2 - 30e + 45.$$

Hence

$$c_1(\hat{X})^2 - 3c_2(\hat{X}) = -e^3(e-5)^2.$$

This is zero if and only if $e = 5$. On the other hand, by Nakai's criterion (see Iitaka [33]), $K_{\hat{X}}$ is ample if $e \geq 3$. Thus the universal covering space of $\hat{X}$ is $\mathbb{B}^2$ if $e = 5$.

NOTE 1.4.11 (1) Yamazaki-Yoshida [67] gave an arithmetic discrete subgroup G of $\mathrm{Aut}(\mathbb{B}^2)$ such that $\mathbb{B}^2/G \simeq \hat{\mathbb{P}}^2$ and $\mathbb{B}^2/[G,G] \simeq \hat{X}$ ($\pi_1(\hat{X}) \simeq [G,G]$) for the above $\hat{\mathbb{P}}^2$ and $\hat{X}$.

(2) Ma. Ishida [34] computed the irregularities $q(\hat{X})$ of Hirzebruch's $\hat{X}$. As for the above $\hat{X}$, $q(\hat{X}) = 30$. See also Fukui [17].

(3) See Höfer [32] for more information on the topics in this section.

1.5 Fenchel's problem

Let M be an n-dimensional compact complex manifold and $D_1,\ldots,D_s$ be distinct irreducible hypersurfaces of M. Let $e_1,\ldots,e_s$ be positive integers greater than 1. Put as before

$$B = D_1 \cup \ldots \cup D_s$$

$$D = e_1D_1 + \ldots + e_sD_s.$$

By Theorem 1.3.9, there is a finite Galois covering

$$\pi : X \longrightarrow M$$

which branches at D if and only if there is a normal subgroup K of $\pi_1(M-B, p_o)$ of finite index with $J \subset K$, which satisfies Condition 1.3.3. However, it is difficult, in general, to look for such a K.

Fenchel's Problem. Give a *good* condition on the pair (M,D) for the existence of a finite Galois covering $\pi : X \longrightarrow M$ which branches at D.

Here a *good* condition means a condition without referring to the fundamental group $\pi_1(M-B, p_o)$.

For a compact Riemann surface M, Theorem 1.2.15 gives an answer to this problem.

If $M = \mathbb{P}^2$ and B is a nodal curve on $\mathbb{P}^2$, then Theorem 1.3.34 gives an answer to this problem. (Note that $\tilde{\pi} : \widetilde{\mathbb{P}}^2(D) \longrightarrow \mathbb{P}^2$ in Theorem 1.3.34 is a finite covering.)

Recently, Kato [38] gave a nice theorem on this problem for a special case:

Let $M = \mathbb{P}^2$ and $D_1,\ldots,D_s$ be lines on $\mathbb{P}^2$. Put

$$\Delta = \{p \in B = D_1 \cup \ldots \cup D_s \mid m_p(B) \geq 3\},$$

where $m_p(B)$ is the multiplicity of the curve B at p.

THEOREM 1.5.1 (Kato). Suppose that

$$D_j \cap \Delta \neq \emptyset \qquad \text{for } 1 \leq j \leq s.$$

Then, for any positive integers $e_1,\ldots,e_s$ greater than 1, there is a finite Galois covering $\pi : X \longrightarrow \mathbb{P}^2$ which branches at $D = e_1D_1 + \ldots + e_sD_s$.

Thus this theorem gives a *good* sufficient condition for the existence in this special case.

This theorem is a special case of

THEOREM 1.5.2 Let M be an n-dimensional projective manifold with $n \geq 2$ and $D_1,\ldots,D_s$ be distinct irreducible hypersurfaces of M. Suppose that there are fixed component free linear pencils $\Lambda_1,\ldots,\Lambda_l$ on M such that (1) every D_j is a member of some Λ_k and (2) every Λ_k contains at least three D_j's as its members. Then, for any positive integers $e_1,\ldots,e_s$ greater than 1, there is a finite Galois covering

$$\pi : X \longrightarrow M$$

which branches at $D = e_1D_1+\ldots+e_sD_s$.

In fact, writing $\Delta = \{p_1,\ldots,p_t\}$ in Theorem 1.5.1, let Λ_k be the linear pencil associated with the projection

$$\pi_{p_k} : \mathbb{P}^2 \longrightarrow \mathbb{P}^1 \qquad \text{(a meromorphic map)}$$

with the centre p_k. Then $\Lambda_1,\ldots,\Lambda_t$ satisfy the conditions of Theorem 1.5.2. Hence Theorem 1.5.1 follows from Theorem 1.5.2.

We now give a proof of Theorem 1.5.2.

First of all, let

$$\pi_k : X_k \longrightarrow M \qquad (1 \leq k \leq r)$$

be finite Galois coverings of M. Let S be an irreducible hypersurface of M at which the ramification index of π_k is a_k (≥ 1). For a *general* point $p \in S$, let $(w_1,\dots,w_n)$ be a local coordinate system around p in M such that $p = (0,\dots,0)$ and $S = \{w_n = 0\}$. For a point $p_k \in \pi_k^{-1}(p)$, let $(x_{k1},\dots,x_{kn})$ be a local coordinate system around p_k such that $p_k = (0,\dots,0)$ and π_k is locally given by

$$\pi_k : (x_{k1},\dots,x_{kn}) \longmapsto (w_1,\dots,w_n) = (x_{k1},\dots,x_{k(n-1)}, x_{kn}^{a_k})$$

(see Theorem 1.1.8). Consider the fibre product and the projection

$$\pi_o : Y = X_1 \times_M \cdots \times_M X_r \longrightarrow M.$$

Then Y is locally given by

$$Y = \{((x_{11},\dots,x_{1n}),\dots,(x_{r1},\dots,x_{rn}))$$

$$x_{k\nu} = x_{j\nu} \quad \text{for} \quad 1 \leq \nu \leq n-1 \quad \text{and} \quad 1 \leq k,j \leq r$$

and

$$x_{kn}^{a_k} = x_{jn}^{a_j} \quad \text{for} \quad 1 \leq k,j \leq r\} .$$

Hence Y can be identified with

$$\{(x_{11},\dots,x_{1n},x_{2n},\dots,x_{rn}) \mid x_{kn}^{a_k} = x_{1n}^{a_1} \quad \text{for } 2 \leq k \leq r\} .$$

Under the identification, π_o is given by

$$\pi_o : (x_{11},\dots,x_{1n},x_{2n},\dots,x_{rn}) \longmapsto (x_{11},\dots,x_{1(n-1)},x_{1n}^{a_1}).$$

In particular, if $a_1 = \dots = a_r = 1$, then π_o is clearly unramified around $(p_1,\dots,p_r)$.

It should be noted that Y may not be irreducible.

The product group

$$G_1 \times \dots \times G_r,$$

where $G_k = G_{\pi_k}$, can be regarded as a group of automorphisms of Y over M. In fact, the action of $G_1 \times \dots \times G_r$ on Y is defined by

$$((\sigma_1,\dots,\sigma_r), (q_1,\dots,q_r)) \in G_1 \times \dots \times G_r \times Y$$

$$\longmapsto (\sigma_1(q_1),\dots,\sigma_r(q_r)) \in Y.$$

Since every π_k is a Galois covering, $G_1 \times \dots \times G_r$ acts transitively on every fibre of π_o. Hence, for any irreducible components X' and X'' of Y, there is $(\sigma_1,\dots,\sigma_r) \in G_1 \times \dots \times G_r$ such that $(\sigma_1,\dots,\sigma_r)(X') = X''$. Moreover,

$$\pi_o : X' \longrightarrow M$$

is a surjective finite map. Let

$$\varphi : X \longrightarrow X'$$

be the normalization of X'. Then

$$\pi = \pi_o \cdot \varphi : X \longrightarrow M$$

is a finite Galois covering such that G_π can be identified with

$$\{(\sigma_1,\dots,\sigma_r) \in G_1 \times \dots \times G_r \mid (\sigma_1,\dots,\sigma_r)(X') = X'\} .$$

If $a_1 = \dots = a_r = a$, then X and π can be locally given by

$$X = \{(x_{11},\dots,x_{1n},x_{2n},\dots,x_{rn}) \mid$$
$$x_{kn} = \zeta^{\nu_k} x_{1n} \quad \text{for} \quad 0 \le \nu_k \le a-1, \quad 2 \le k \le r\},$$

$$\pi : (x_{11},\dots,x_{1n},x_{2n},\dots,x_{rn}) \longrightarrow (x_{11},\dots,x_{1(n-1)},x_{1n}^a),$$

where $\zeta = \exp 2\pi\sqrt{-1}/a$, and $\nu_2,\dots,\nu_s$ are fixed. Thus

<u>LEMMA 1.5.3</u> Under the notations as above, if $a_1 = \dots = a_r = a$, then the ramification index of π at S is a.

Now, we are ready to prove Theorem 1.5.2. Under the notations and conditions of Theorem 1.5.2, by the fundamental theorem of Hironaka [28], there is a bimeromorphic (i.e. a birational) holomorphic map

$$\rho : \hat{M} \longrightarrow M$$

such that (1) $\hat{M}$ is a projective manifold, (2) $\rho : M - \rho^{-1}(B) \longrightarrow M - B$ is biholomorphic, where $B = D_1 \cup \dots \cup D_s$, (3) $\rho^{-1}(B)$ is a simple normally crossing hypersurface of M and (4) the linear pencil $\tilde{\Lambda}_k$ on $\hat{M}$ induced from Λ_k is fixed component free and base point free for $1 \le k \le t$.

Let $\hat{D}_j$ be the strict transform of D_j. Then, by the conditions of Theorem 1.5.2,

(1)' every $\hat{D}_j$ is an irreducible component of a member of some $\tilde{\Lambda}_k$ and

(2)' every $\tilde{\Lambda}_k$ contains at least three members which have some $\hat{D}_j$'s as irreducible components.

Let

$$\{\hat{D}_1, \ldots, \hat{D}_m\} \qquad (m \geq 3)$$

be the set of all $\hat{D}_j$'s which are irreducible components of members of $\tilde{\Lambda}_1$, say. Such members can be written as

$$\hat{D}_j + \hat{F}_j \quad \text{for} \quad 1 \leq j \leq m,$$

where $\hat{F}_j$ is either the zero divisor or an exceptional divisor of $\rho : \hat{M} \longrightarrow M$. (A positive divisor $\hat{F}$ on $\hat{M}$ is called an *exceptional divisor of* $\rho : \hat{M} \longrightarrow M$ if

$$\dim \rho(\mathrm{Supp}(\hat{F})) \leq n-2,$$

where $\mathrm{Supp}(\hat{F})$ is the support of $\hat{F}$.) Such an exceptional divisor $\hat{F}_j$ may exist, because D_j may be a member of other Λ_k's.

Let

$$\hat{f}_1 : \hat{M} \longrightarrow \mathbb{P}^1$$

be the holomorphic map associated with $\tilde{\Lambda}_1$. Put

$$\lambda_j \doteq \hat{f}_1(\hat{D}_j + \hat{F}_j) \in \mathbb{P}^1 .$$

By (2)' and Theorem 1.2.15, there is a finite Galois covering

$$\alpha : R \longrightarrow \mathbb{P}^1$$

which branches at $e_1\lambda_1 + \ldots + e_m\lambda_m$. Consider the fibre product

$$\beta : \hat{Y}_1 = \hat{M} \times_{\mathbb{P}^1} R \longrightarrow \hat{M}.$$

Note that there is a finite set Γ in $\mathbb{P}^1$ such that

$$\hat{f}_1 : \hat{M} - \hat{f}_1^{-1}(\Gamma) \longrightarrow \mathbb{P}^1 - \Gamma$$

is of maximal rank at every point. Hence this is a C^∞-bundle over $\mathbb{P}^1 - \Gamma$. Its fibre is connected, since a general member of Λ_1 is irreducible. Hence every fibre of

$$\hat{Y}_1 \longrightarrow R$$

is irreducible. Hence $\hat{Y}_1$ is itself irreducible. Let

$$\varphi : \hat{X}_1 \longrightarrow \hat{Y}_1$$

be the normalization of $\hat{Y}_1$. Then

LEMMA 1.5.4 The composition

$$\hat{\pi}_1 = \beta.\varphi : X_1 \longrightarrow M$$

is a finite Galois covering which branches at

$$e_1\hat{D}_1 + \ldots + e_m\hat{D}_m + \hat{F}'$$

such that $G_{\hat{\pi}_1} \simeq G_\alpha$, where $\hat{F}'$ is either the zero divisor or an exceptional divisor of $\rho : \hat{M} \longrightarrow M$.

PROOF. Let $\hat{p}$ be a *general* point of $\hat{D}_1$, say. Put $\lambda = \hat{f}_1(\hat{p})$ and take a point $q \in \alpha^{-1}(\lambda)$. Let $(x_1,\ldots,x_n)$, w and z be local coordinate systems around $\hat{p},\lambda$ and q, respectively, such that $\hat{f}_1$ and α are locally given by

$$\hat{f}_1 : (x_1,\ldots,x_n) \longmapsto w = x_n,$$
$$\alpha : z \longmapsto w = z^{e_1}.$$

Then $\hat{X}_1$ and $\hat{\pi}_1$ are locally given by

$$\hat{X}_1 = \{(x_1,\ldots,x_n,z) \mid x_n = z^{e_1}\},$$
$$\hat{\pi}_1 : (x_1,\ldots,x_n,z) \longmapsto (x_1,\ldots,x_n).$$

Hence, the ramification index of $\hat{\pi}_1$ at $\hat{D}_1$ is e_1.

A similar argument shows that the ramification index of $\hat{\pi}_1$ at a prime divisor $\hat{E}$ contained in $\hat{F}_1$ with the multiplicity b (i.e. $\hat{F}_1 = b\hat{E} + *$) is $e_1/(e_1,b)$.

The other assertions of the lemma are easy to show. □

In a similar way, we get a finite Galois coverings

$$\hat{\pi}_k : \hat{X}_k \longrightarrow \hat{M} \quad \text{for} \quad 2 \leq k \leq t.$$

Let

$$\hat{\pi}_o : \hat{Y} = \hat{X}_1 \times_{\hat{M}} \cdots \times_{\hat{M}} \hat{X}_t \longrightarrow \hat{M}$$

be the fibre product. Let $\hat{X}'$ be an irreducible component of $\hat{Y}$ and

$$\hat{\varphi} : \hat{X} \longrightarrow \hat{X}'$$

be the normalization of $\hat{X}'$. Then, by Lemma 1.5.3,

$$\hat{\pi} = \hat{\pi}_o \cdot \hat{\varphi} : \hat{X} \longrightarrow \hat{M}$$

is a finite Galois covering which branches at

$$e_1\hat{D}_1 + \dots + e_s\hat{D}_s + \hat{F},$$

where $\hat{F}$ is either the zero divisor or an exceptional divisor of $\rho : \hat{M} \longrightarrow M$.

Now, let

$$\pi : X \longrightarrow M$$

be the $\mathbb{C}(\hat{X})$-normalization of M. Then π is clearly a finite Galois covering which branches at

$$D = e_1D_1 + \dots + e_sD_s,$$

such that $G_\pi \simeq G_{\hat{\pi}}$. This completes the proof of Theorem 1.5.2.

REMARK 1.5.5 (1) As is easily seen, Theorem 1.5.2 still holds even if $D_1,\dots,D_s$ are reduced hypersurfaces, any two of which do not have a common irreducible component.
(2) Kato's original proof ([38]) of Theorem 1.5.1 also uses Theorem 1.2.15, together with *Zariski pencils* (see Zariski [69]).

COROLLARY 1.5.6 Let M be a non-singular irreducible projective surface and $D_1,\dots,D_s$ be distinct non-singular irreducible curves on M such that $B = D_1 \cap \dots \cap D_s$ has only ordinary multiple points as its singularity. For a point $p \in \mathrm{Sing}B$, put

$$A(p) = \{D_j \mid p \in D_j\}.$$

Put also

$$\Delta = \{q \in B \mid m_q(B) \geq 3\}.$$

Suppose that (1) $D_j \cap \Delta \neq \emptyset$ for $1 \leq j \leq s$, (2) curves in A(p) are linearly equivalent for every $p \in \Delta$, and (3) for any points p and q in SingB, one of the following holds: (a) A(p) = A(q), (b) $A(p) \cap A(q)$ = a unique curve D_j, (c) $A(p) \cap A(q) = \emptyset$. Then, for any positive integers $e_1,\dots,e_s$ greater than 1, there is a finite Galois covering $\pi : X \longrightarrow M$ which branches at $D = e_1D_1 + \dots + e_sD_s$.

PROOF For a point $p \in \Delta$, we may put, say,

$$A(p) = \{D_1, \ldots, D_m\}, \qquad (m \geq 3).$$

By the condition (2), $D_1, \ldots, D_m$ are linearly equivalent. We show that they all belong to a linear pencil Λ_p. It suffices to show that every D_j $(3 \leq j \leq m)$ belongs to the linear pencil Λ_p generated by D_1 and D_2. We write

$$D_1 = (\xi_1) \quad \text{and} \quad D_2 = (\xi_2),$$

where ξ_1 and ξ_2 are holomorphic sections of the line bundle $[D_1] = [D_2]$. Then

$$\Lambda_p = \{D_\lambda = (\lambda_1\xi_1 + \lambda_2\xi_2) \mid \lambda = (\lambda_1 : \lambda_2) \in \mathbb{P}^1\} .$$

Take a point $r \in D_3$ such that $r \notin D_1 \cup D_2$. Then, since $\xi_1(r) \neq 0$ and $\xi_2(r) \neq 0$, there is a point $\mu = (\mu_1 : \mu_2) \in \mathbb{P}^1$ with $\mu_1 \neq 0$ and $\mu_2 \neq 0$ such that

$$\mu_1\xi_1(r) + \mu_2\xi_2(r) = 0.$$

Since $D_3 \sim D_\mu \sim D_1$, the intersection numbers satisfy

$$D_3 D_\mu = D_1^2 .$$

If D_3 and D_μ do not have a common irreducible component, then, by the condition (3),

$$D_3 \cap D_\mu \supset (D_1 \cap D_2) \cup [r].$$

Hence

$$D_3 D_\mu \geq D_1 D_2 + 1 = D_1^2 + 1,$$

a contradiction. Hence D_3 and D_μ have a common irreducible component. Since D_3 is itself irreducible, $D_3 \leq D_\mu$. But $D_3 \sim D_\mu$. Hence

$$D_3 = D_\mu \in \Lambda_p.$$

In a similar way, $D_4, \ldots, D_m$ belong to Λ_p.

Now, by condition (1),

$$\{\Lambda_p \mid p \in \Delta\}$$

satisfy the conditions of Thereom 1.5.2. □

REMARK 1.5.7 (1) For p and q in Λ, $\Lambda_p = \Lambda_q$ if and only if $q \in Bs\Lambda_p$, the base locus of Λ_p. (2) This corollary clearly implies Theorem 1.5.1. (3) In this corollary, let

$$\rho : \hat{M} \longrightarrow M$$

be the blowing up of M at all points of Δ. For a point $p \in \Delta$, let

$$A'(p) = \{D_1, \ldots, D_h\}, \quad \text{(say)},$$

be the set of all $D_j \in A(p)$ such that $D_j \in A(q)$ for some point $q \in \Delta$ with $q \notin Bs\Lambda_p$. Put

$$e_p \begin{cases} = \langle e_1, \ldots, e_h \rangle, & \text{(the LCM)} \\ = 0, & \text{if } A'(p) \text{ is empty.} \end{cases}$$

Then the finite Galois covering

$$\hat{\pi} : \hat{X} \longrightarrow \hat{M}$$

in the proof of Theorem 1.5.2 branches, in this case, at

$$e_1\hat{D}_1 + \ldots + e_s\hat{D}_s + \sum_{p \in \Delta} e_pE_p,$$

where $\hat{D}_j$ is the strict transform of D_j and E_p is the exceptional curve over $p \in \Delta$.

The proof of Theorem 1.5.2 can be applied to prove the following theorem:

THEOREM 1.5.8 Let $C_1, \ldots, C_s$ $(s \geq 2)$ be irreducible conics in $\mathbb{P}^2$. Assume that, for every C_j, there is another C_k such that C_j and C_k are tangent at two distinct points. Then, for any positive integers $e_1, \ldots, e_s$ greater than 1, there is a finite Galois covering $\pi : X \longrightarrow \mathbb{P}^2$ which branches at $D = e_1C_1 + \ldots + e_sC_s$.

PROOF. If 3 or more conics $C_1, \ldots, C_k$ $(k \geq 3)$, say, are tangent at the same points p and q $(p \neq q)$, (see the mouth of the face below), then we consider the linear pencil Λ to which $C_1, \ldots, C_k$ belong as members.

If only 2 conics C_ℓ and C_m, say, are tangent at points p and q ($p \neq q$), (see the eyes of the above face), then we consider the line L passing through p and q.

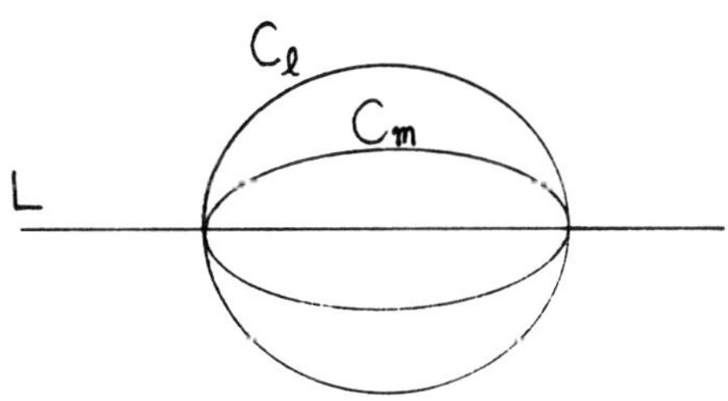

2L is then a member of the linear pencil Λ' containing C_ℓ and C_m. Consider a bimeromorphic holomorphic map

$$\rho : \hat{M} \longrightarrow M$$

which satisfies the conditions (1)-(3) in the proof of Theorem 1.5.2. The linear pencil $\tilde{\Lambda}'$ on $\hat{M}$ induced from Λ' is base point free, so

$$\Phi = \Phi_{\tilde{\Lambda}'} : \hat{M} \longrightarrow \mathbb{P}^1$$

is a holomorphic map. Let $\hat{C}_\ell, \hat{C}_m, \hat{L}$ be the strict transforms of C_ℓ, C_m, L by ρ, respectively. Put

$$\lambda_\ell = \Phi(\hat{C}_\ell), \ \lambda_m = \Phi(\hat{C}_m) \quad \text{and} \quad \lambda = \Phi(L).$$

Let

$$\alpha : R \longrightarrow \mathbb{P}^1$$

is a finite Galois covering which branches at $e_\ell\lambda_\ell + e_m\lambda_m + 2\lambda$. Consider the fibre product $\hat{Y} = \hat{M} \times_{\mathbb{P}^1} R$ and its normalization $\hat{X}$:

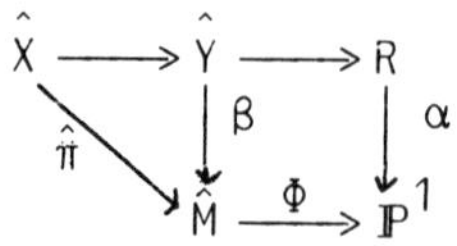

Then, as is easily seen, $\hat{\pi} : \hat{X} \longrightarrow \hat{M}$ is a finite Galois covering which branches at

$$e_\ell\hat{C}_\ell + e_m\hat{C}_m + \hat{L} + \hat{F} \qquad (\hat{F} = \text{an exceptional divisor}),$$

that is, *does not* branch at $\hat{L}$, so branches at

$$e_\ell\hat{C}_\ell + e_m\hat{C}_m + \hat{F}.$$

The other parts of the proof are similar to those of Theorem 1.5.2. □

This kind of method can be applied also to prove, for example,

<u>PROPOSITION 1.5.9</u> Let L_1, L_2, L_3 be 3 lines in $\mathbb{P}^2$ circumscribing an irreducible conic C. Then, for any positive integers a and b greater than 1, there is a finite Galois covering $\pi : X \longrightarrow \mathbb{P}^2$ which branches at $a(L_1+L_2+L_3) + bC$.

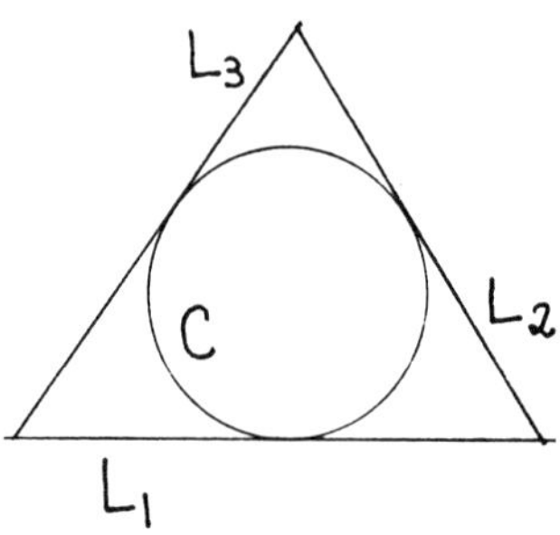

Finally, we state the following theorem:

THEOREM 1.5.10 Let M be a projective manifold and $D_1,\ldots,D_s$ be be irreducible hypersurfaces of M. Put $B = D \cup \ldots \cup D_s$. Let a non-trivial representation of $\pi_1(M-B,\ p_o)$:

$$R : \pi_1(M-B,\ p_o) \longrightarrow GL(r,\ \mathbb{C})$$

be given. Let γ_j be a loop in $M-B$ which rounds D_j-SingB once in a counter-clockwise direction. Suppose that the orders of $R(\gamma_1),\ldots,R(\gamma_s)$ are finite. Call them $e_1,\ldots,e_s$, respectively. Then there is a finite Galois covering $\pi : X \longrightarrow M$ which branches at $D = e_1D_1+\ldots+e_sD_s$.

Though the condition of the theorem is not *good*, it seems worth stating. This theorem is an easy consequence of the following two theorems:

THEOREM 1.5.11 $\pi_1(M-B,\ p_o)$ is finitely generated.

THEOREM 1.5.12 (Selberg [54]). Let Γ be a (non-trivial) finitely generated subgroup of $GL(n,\ \mathbb{C})$. Then there is a torsion free normal subgroup N ($\neq \Gamma$) of Γ of finite index.

Theorem 1.5.11 is due to Professor M. Oka. His proof uses a method similar to that in Hamm-Lê [27], but it is too long to state here.

2 Fields of algebraic functions

2.1 Galois theory of fields of algebraic functions

Henceforth, we consider the following situation: Let $\bar{M}$ be an n-dimensional projective manifold and $\bar{B}$ be a hypersurface of $\bar{M}$. ($\bar{B}$ may possibly be empty.) Put

$$M = \bar{M} - \mathrm{Sing}\bar{B} \text{ and } B = \bar{B} - \mathrm{Sing}\bar{B}.$$

Then M is a quasi-projective manifold and B is a non-singular hypersurface of M. Let

$$\bar{B} = \bar{D}_1 \cup \ldots \cup \bar{D}_s$$

be the irreducible decomposition of $\bar{B}$. Then

$$B = D_1 \cup \ldots \cup D_s, \quad (D_j = \bar{D}_j - \mathrm{Sing}\bar{B})$$

is the decomposition of B into connected components. Let $e_1,\ldots,e_s$ be positive integers. Put

$$\bar{D} = e_1\bar{D}_1 + \ldots + e_s\bar{D}_s$$

and

$$D = e_1D_1 + \ldots + e_sD_s.$$

We suppose that Condition 1.2.4 holds on the pair (M,D). Then, by Theorem 1.2.7, there is a maximal covering

$$\tilde{\pi} : \tilde{M} = \tilde{M}(D) \longrightarrow M$$

(of non-singular type) which branches at D. This is unique up to equivalence.

For a meromorphic function $f \in \mathbb{C}(M)$ on M,

$$\tilde{\pi}^*(f) = f.\tilde{\pi}$$

is the pullback of f by $\tilde{\pi}$. This is locally defined by

$$\tilde{\pi}^*(f) = f.\tilde{\pi} = f_1.\tilde{\pi}/f_2.\tilde{\pi} \quad \text{if } f = f_1/f_2,$$

where f_1 and f_2 are holomorphic functions on an open set of M. $\tilde{\pi}^*(f)$ is then a meromorphic function on $\tilde{M}$ and

$$\tilde{\pi}^* : \mathbb{C}(M) \longrightarrow \mathbb{C}(\tilde{M})$$

is an injective homomorphism. We regard $\mathbb{C}(M)$ as a subfield of $\mathbb{C}(\tilde{M})$ through $\tilde{\pi}^*$.

In general, if $\pi : X \longrightarrow M$ is a branched covering, then π induces an injective homomorphism

$$\begin{array}{cccc} \pi^* : & \mathbb{C}(M) & \longrightarrow & \mathbb{C}(X). \\ & \Psi & & \Psi \\ & f & \longmapsto & f.\pi \end{array}$$

We regard $\mathbb{C}(M)$ as a subfield of $\mathbb{C}(X)$ through π^*.

Moreover, a covering transformation $\varphi \in G$ induces an automorphism

$$\begin{array}{cccc} \varphi^* : & \mathbb{C}(X) & \longrightarrow & \mathbb{C}(X) \\ & \Psi & & \Psi \\ & h & \longmapsto & h.\varphi \end{array}$$

which satisfies

$$\varphi^*(\pi^*(f)) = \pi^*(f) \quad \text{for all } f \in \mathbb{C}(M).$$

<u>LEMMA 2.1.1</u> If $\pi : X \longrightarrow M$ is a Galois covering which branches at most at D, then

$$\mathbb{C}(M) = \{h \in \mathbb{C}(X) \mid \varphi^*(h) = h \quad \text{for all } \varphi \in G_\pi\} .$$

In particular,

$$\mathbb{C}(M) = \{h \in \mathbb{C}(\tilde{M}) \mid \varphi^*(h) = h \quad \text{for all } \varphi \in G_{\tilde{\pi}}\} .$$

<u>PROOF</u>. For $h \in \mathbb{C}(X)$, suppose that

$$\varphi^*(h) = h \quad \text{for all } \varphi \in G_\pi .$$

Since $\pi : X - \pi^{-1}(B) \longrightarrow M - B$ is an unramified Galois covering, there is a meromorphic function f on M-B such that

$$\pi^*f = h \quad \text{on} \quad X - \pi^{-1}(B).$$

We show that f can be extended to a meromorphic function on M. Take a point $p \in B$ and a local coordinate system $(w_1,\ldots,w_n)$ around p such that $p = (0,\ldots,0)$ and $B = \{w_n = 0\}$ around p. Take a point $q \in \pi^{-1}(p)$. By Theorem 1.1.8, there is a local coordinate system $(z_1,\ldots,z_n)$ around q such that $q = (0,\ldots,0)$, $\pi^{-1}(B) = \{z_n = 0\}$ around q and π is locally given by

$$\pi : (z_1,\dots,z_n) \longmapsto (w_1,\dots,w_n) = (z_1,\dots,z_{n-1},z_n^e)$$

for a positive integer e. Consider the local holomorphic isomorphism

$$\sigma : (z_1,\dots,z_n) \longmapsto (z_1,\dots,z_{n-1}, \zeta z_n)$$

around q, where $\zeta = \exp(2\pi\sqrt{-1}/e)$. Then $\pi\sigma = \pi$. Since π is a Galois covering, σ can be uniquely extended to a covering transformation $\sigma \in G_\pi$.

Suppose that q is not a point of $D_\infty(h) - \{z_n = 0\}$. Then h can be expanded into a Laurent series as follows:

$$h(z_1,\dots,z_n) = a_m(z')z_n^m + a_{m+1}(z')z_n^{m+1} + \dots ,$$

where m is an integer and $a_m(z'), a_{m+1}(z'),\dots$ are holomorphic functions of $z' = (z_1,\dots,z_{n-1})$ such that $a_m(z')$ is not identically zero. Since $\sigma^*h = h$, we have easily

$$a_\nu(z') = 0 \quad \text{(identically)} \quad \text{if } \nu \not\equiv 0 \pmod e.$$

Hence m = ke for an integer k and h can be written as

$$h(z_1,\dots,z_n) = a_{ke}(z')z_n^{ke} + a_{(k+1)e}(z')z_n^{(k+1)e} + \dots .$$

Consider the meromorphic function

$$f(w_1,\dots,w_n) = a_{ke}(w')w_n^k + a_{(k+1)e}(w')w_n^{k+1} + \dots$$

around p, where $w' = (w_1,\dots,w_{n-1})$. Then $\pi^*f = h$. Hence f on M-B can be extended to a meromorphic function around p.

The closure $\overline{D_\infty(h) - \pi^{-1}(B)}$ of $D_\infty(h) - \pi^{-1}(B)$ is a hypersurface of X which is G_π-invariant. By Proposition 1.1.11, $\pi(\overline{D_\infty(h) - \pi^{-1}(B)})$ is a closed complex subspace of M.

$$S = \pi(\overline{D_\infty(h) - \pi^{-1}(B)}) \cap B$$

is then a closed complex subspace of M of codimension greater than 1. f has been already extended to M-S. By Levi's theorem (see Narasimhan [45]), f can be extended to a meromorphic function f on M, which satisfies $\pi^*f = h$. □

We note that, by Levi's theorem,

$$\mathbb{C}(M) = \mathbb{C}(\bar{M}).$$

LEMMA 2.1.2 Let $\pi : X \longrightarrow M$ be a finite covering which branches at most at D. Then X is a quasi-projective manifold.

PROOF. By Grauert-Remmert's theorem ([21]), X is an algebraic variety such that the field $\mathbb{C}(X)_{alg}$ of rational functions on X is a finite extension of $\mathbb{C}(M) = \mathbb{C}(\bar{M})$. Let

$$\bar{\pi} : \bar{X} \longrightarrow \bar{M}$$

be the $\mathbb{C}(X)_{alg}$-normalization of $\bar{M}$. Then it is clear that X can be identified with the Zariski open set of $\bar{X}$ such that $\bar{X} - X = \bar{\pi}^{-1}(\mathrm{Sing}\bar{B})$. By Theorem 1.3.24, $\bar{X}$ is a projective variety. Hence X is a quasi-projective manifold. □

For a subgroup H of $G_{\tilde{\pi}}$, put

$$K(H) = \{h \in \mathbb{C}(\tilde{M}) \mid \varphi^*(h) = h \text{ for all } \varphi \in H\}.$$

Then K(H) is a subfield of $\mathbb{C}(\tilde{M})$ such that $\mathbb{C}(M) \subset K(H)$. Note that, if $H_1 \subset H_2$, then $K(H_1) \supset K(H_2)$.

For a subfield K of $\mathbb{C}(\tilde{M})$ such that $\mathbb{C}(M) \subset K$, put

$$G(K) = \{\varphi \in G_{\tilde{\pi}} \mid \varphi^*(h) = h \text{ for all } h \in K\}.$$

Then G(K) is a subgroup of $G_{\tilde{\pi}}$. Note that, if $K_1 \subset K_2$, then $G(K_1) \supset G(K_2)$.

THEOREM 2.1.3 (1) If H is a subgroup of $G_{\tilde{\pi}}$ of finite index, then $G(K(H)) = H$ and $[K(H) : \mathbb{C}(M)] = [G:H]$. (2) If K is a subfield of $\mathbb{C}(\tilde{M})$ such that $\mathbb{C}(M) \subset K$, and $K/\mathbb{C}(M)$ is a finite extension, then $K(G(K)) = K$ and $[G_{\tilde{\pi}} : G(K)] = [K : \mathbb{C}(M)]$.
(3) The correspondences $H \longmapsto K(H)$ and $K \longmapsto G(K)$ in (1) and (2) are one-to-one correspondences. Moreover, $\mathbb{C}(\tilde{M})$ contains all conjugate subfields of $K/\mathbb{C}(M)$, which correspond to conjugate subgroups of $H = G(K)$.
(4) For a subfield K of $\mathbb{C}(\tilde{M})$ such that $\mathbb{C}(M) \subset K$, $K/\mathbb{C}(M)$ is a finite Galois extension if and only if G(K) is a normal subgroup of $G_{\tilde{\pi}}$ of finite index. In this case, the quotient group $G_{\tilde{\pi}}/G(K)$ is naturally isomorphic to the Galois group $\mathrm{Gal}(K/\mathbb{C}(M))$ of $K/\mathbb{C}(M)$.

PROOF (1). For a subgroup H of $G_{\tilde{\pi}}$, there is a corresponding equivalence class of coverings $\pi : X \longrightarrow M$ which branches at most at D, (see Theorem 1.2.7):

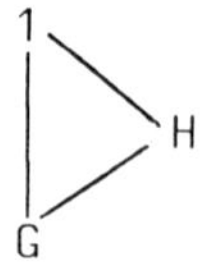

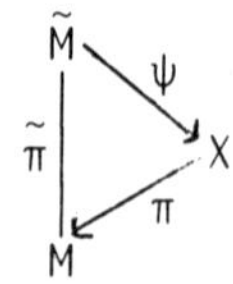

A morphism $\psi : \tilde{M} \longrightarrow X$ of $\tilde{\pi}$ to π is clearly a Galois covering with $G_\psi = H$. Hence, by Lemma 2.1.1,

$$\mathbb{C}(X) = K(H).$$

It is clear that

$$H \subset G(K(H)).$$

Take any element $\varphi \in G_{\tilde{\pi}}-H$. Take a point $p \in \tilde{M}-\tilde{\pi}^{-1}(B)$. Then

$$\psi(\varphi(p)) \neq \psi(p).$$

In fact, if $\psi(\varphi(p)) = \psi(p)$, then there is a unique $\sigma \in H$ such that $\varphi(p) = \sigma(p)$, because $\psi : \tilde{M} \longrightarrow X$ is a Galois covering. Hence $\varphi = \sigma \in H$, a contradiction.

Now, if H is of finite index in $G_{\tilde{\pi}}$, then

$$+\infty > [G_{\tilde{\pi}}:H] = \deg \pi = [\mathbb{C}(X) : \mathbb{C}(M)] = [K(H) : \mathbb{C}(M)].$$

By Lemma 2.1.2, X is a quasi-projective manifold. Hence there is $h' \in \mathbb{C}(X) = K(H)$ such that

$$h'(\psi(\varphi(p))) \neq h'(\psi(p)).$$

Put

$$h = \psi^*(h') = h'.\psi \in \mathbb{C}(\tilde{M}).$$

Then

$$h(\varphi(p)) \neq h(p).$$

This implies that

$$\varphi^*(h) \neq h.$$

Since $h \in K(H)$, we have

$$\varphi \notin G(K(H)).$$

This holds for any $\varphi \in G_{\tilde{\pi}}-H$. Thus

$H = G(K(H))$.

(2). We have clearly

$K \subset K(G(K))$.

Since $K/\mathbb{C}(M)$ is a simple extension, there is an element $h \in K$ such $K = \mathbb{C}(M)(h)$. Let

$F(x) = 0$

be an irreducible equation over $\mathbb{C}(M)$ which h satisfies. Let

$G_{\tilde{\pi}} = \cup G(K)\varphi_j$

be the coset decomposition. Then

$F(\varphi_j^*(h)) = \varphi_j^*(F(h)) = 0$.

Note that

$\varphi_j^*(h) \neq \varphi_k^*(h)$ for $j \neq k$.

In fact, if $\varphi_j^*(h) = \varphi_k^*(h)$ for some j and k with $j \neq k$, then $(\varphi_j\varphi_k^{-1})^*(h) = h$. Hence $(\varphi_j\varphi_k^{-1})^*(h') = h'$ for all $h' \in \mathbb{C}(M)(h) = K$. Hence $\varphi_j\varphi_k^{-1} \in G(K)$, a contradiction.

Thus we have

$$\infty > [K : \mathbb{C}(M)] = \deg F \geq [G_{\tilde{\pi}} : G(K)] \geq [G_{\tilde{\pi}} : G(K(G(K)))].$$

Hence, by (1),

$G(K(G(K))) = G(K)$.

Hence, by (1) again,

$$[K : \mathbb{C}(M)] = \deg F \geq [G_{\tilde{\pi}} : G(K)] = [G_{\tilde{\pi}} : G(K(G(K)))]$$
$$= [K(G(K)) : \mathbb{C}(M)] \geq [K : \mathbb{C}(M)].$$

Hence, they are all equal. In particular

$K(G(K)) = K$.

(3). This follows from the proof of (2).

(4). This follows from (3). □

<u>REMARK 2.1.4</u> We do not know if (1) of Theorem 2.1.3 still holds even if H is of infintie index in $G_{\tilde{\pi}}$. On the other hand, at the end of the

next section, we will give an example of fields L with $\mathbb{C}(M) \subset L \subset \mathbb{C}(\tilde{M})$ such that $[L : \mathbb{C}(M)] = +\infty$ and $K(G(L)) \neq L$.

2.2 Representative matrix functions

We follow the notations and the assumptions in Section 2.1. Let r be a positive integer and

$$R : G_{\tilde{\pi}} \longrightarrow GL(r,\mathbb{C})$$

be a homorphism, that is, a representation of $G_{\tilde{\pi}}$. An (r × r)-matrix (resp. r-vector) valued meromorphic function $F = (f_{jk})$ (resp. $f = {}^t(f_1,\ldots,f_r)$, a column vector) on $\tilde{M} = \tilde{M}(D)$ is called a *representative matrix* (resp. *vector*) *function with respect to* R, if

$$\varphi^*F = R(\varphi)F \quad \text{for all } \varphi \in G_{\tilde{\pi}},$$

$$(\text{resp. } \varphi^*f = R(\varphi)f \quad \text{for all } \varphi \in G_{\tilde{\pi}}),$$

where $\varphi^*F = F.\varphi = (\varphi^*f_{jk}) = (f_{jk}.\varphi)$, (resp. $\varphi^*f = f.\varphi = {}^t(\varphi^*f_1,\ldots,\varphi^*f_r) = {}^t(f_1.\varphi,\ldots,f_r.\varphi)$). The functions f_{jk} and f_j are called *representative functions* (*with respect to* R).

Note that the determinant det(F) of a representative matrix function F is a meromorphic function on $\tilde{M}$ such that

$$\varphi^*(\det(F)) = \det(R(\varphi))\det(F) \quad \text{for all } \varphi \in G_{\tilde{\pi}}.$$

Hence det(F) is a representative matrix function (r = 1) with respect to the representation

$$\det(R) : \varphi \in G_{\tilde{\pi}} \longmapsto \det(R(\varphi)) \in \mathbb{C}^*.$$

A representative matrix function F is said to be *non-degenerate* if det(F) is not identically zero.

The following theorem is fundamental in our discussion.

THEOREM 2.2.1 For any representation R of $G_{\tilde{\pi}}$, there is a non-degenerate representative matrix function with respect to R.

For the proof of the theorem, we need some preparations. Put

$$M^* = M - B \quad \text{and} \quad \tilde{M}^* = \tilde{M} - \tilde{\pi}^{-1}(B).$$

First of all, we recall that a representation R induces a flat vector bundle

$$\tau : S_R \longrightarrow M^*.$$

The bundle space S_R is the quotient space

$$S_R = (\tilde{M}^* \times \mathbb{C}^r)/G_{\tilde{\pi}},$$

where $G_{\tilde{\pi}}$ acts on $\tilde{M}^* \times \mathbb{C}^r$ as follows:

$$(\varphi, p, \xi) \in G_{\tilde{\pi}} \times \tilde{M}^* \times \mathbb{C}^r \longmapsto (\varphi(p), R(\varphi)\xi) \in \tilde{M}^* \times \mathbb{C}^r.$$

The projection τ is the one naturally induced by the projection

$$(p, \xi) \in \tilde{M}^* \times \mathbb{C}^r \longmapsto p \in \tilde{M}^*.$$

LEMMA 2.2.2 A global holomorphic (resp. meromorphic) section of the vector bundle $\tau : S_R \longrightarrow M^*$ can be naturally identified with an r-vector valued holomorphic (resp. meromorphic) function f on $\tilde{M}^*$ such that

$$\varphi^* f = R(\varphi) f \quad \text{for all } \varphi \in G_{\tilde{\pi}}.$$

PROOF. Note that the induced bundle $\tilde{\pi}^*(S_R)$ is nothing but the product bundle $\tilde{M}^* \times \mathbb{C}^r \longrightarrow \tilde{M}^*$. Hence a global holomorphic (resp. meromorphic) section ξ of $\tau : S_R \longrightarrow M^*$ induces a global holomorphic (resp. meromorphic) section $\tilde{\xi}$ of $\tilde{M}^* \times \mathbb{C}^r \longrightarrow \tilde{M}^*$, which can be identified with an r-vector valued holomorphic (resp. meromorphic) function f on $\tilde{M}^*$ such that

$$\varphi^* f = R(\varphi) f \quad \text{for all } \varphi \in G_{\tilde{\pi}}.$$

Conversely, it is clear that such a function f on $\tilde{M}^*$ induces a global section of $\tau : S_R \longrightarrow M^*$. □

We observe the identification in Lemma 2.2.2 from another point of view, following Steenrod [59]. As in section 1.2, we choose a point $p_o \in M$ (the base point) and fix it once and for all. Let $\{W_j\}$ be an open covering of M^* such that (1) every W_j and $W_j \cap W_k$ are connected and simply connected and (2) S_R is trivial on W_j for every j. Let W_o be a member of $\{W_j\}$ such that $p_o \in W_o$. Choose a point $p_j \in W_j$ for every j. (Choose the point p_o for W_o.) Let δ_j be a path in M^* which connects from p_o to p_j. For a point $p \in W_j \cap W_k$, let $\delta_{jk}(p)$ (resp. $\delta_{kj}(p)$) be a path in M^* which connects from p_j (resp. p_k) to p. Then $\delta = \delta_j \delta_{jk}(p) \delta_{kj}(p)^{-1} \delta_k^{-1}$ is a closed path in M^* which starts from p_o.

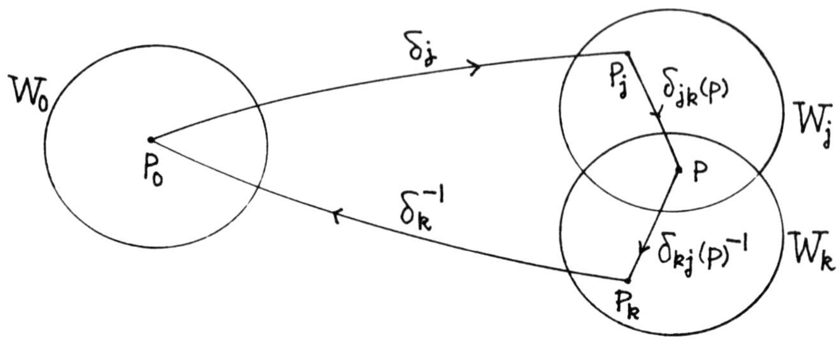

We regard δ as an element of $\pi_1(M^*, p_o)$. Then δ does not clearly depend on $p \in W_j \cap W_k$. Moreover, by the natural homomorphism

$$\pi_1(M^*, p_o) \longrightarrow G_{\tilde{\pi}},$$

we may regard δ as an element of $G_{\tilde{\pi}}$. Put

$$g_{jk} = R(\delta).$$

Then g_{jk} is a constant matrix on $W_j \cap W_k$. The flat vector bundle S_R is simply the one defined by the constant transitions g_{jk} : $S_R = \{g_{jk}\}$.

Let $\xi = \{\xi_j\}$ be a global holomorphic (resp. meromorphic) section of S_R, that is, ξ_j is an r-vector valued holomorphic (resp. meromorphic) function on W_j such that

$$\xi_j = g_{jk}\xi_k \quad \text{on} \quad W_j \cap W_k.$$

The function $f = \xi_o$ is continued analytically along curves γ in M^* which start from p_o. If $\gamma \subset W_o \cup W_1$, then $\xi_o(p) = g_{o1}\cdot\xi_1(p)$ for $p \in W_o \cap W_1$. Hence, we may define

$$f(p) = g_{o1}\cdot\xi_1(p) \quad \text{for } p \in W_1.$$

Thus $f = \xi_o$ is continued on $W_o \cup W_1$. We repeat this process. If $\gamma \subset W_o \cup W_1 \cup \ldots \cup W_k$, then we may define

$$f(p) = g_{o1}\cdots g_{k-1,k}\cdot\xi_k(p) \quad \text{for } p \in W_k.$$

Thus $f = \xi_o$ is continued on $W_o \cup W_1 \cup \ldots \cup W_k$.

In particular, if γ is a closed path, then $f = \xi_o$ is continued on $W_k = W_o$

to the function

$$g_{01}\cdots g_{k-1,k}\xi_k = g_{01}\cdots g_{k-1,k}\xi_o$$
$$= R(\delta_o\delta_{01}(p_1')\delta_{1o}(p_1')^{-1}\delta_1^{-1}\cdot\delta_1\delta_{12}(p_2')\delta_{21}(p_2')^{-1}\delta_2^{-1}$$
$$\cdots\ \delta_{k-1}\delta_{k-1,k}(p_k')\delta_{k,k-1}(p_k')^{-1}\delta_k^{-1})\xi_o$$
$$= R(\gamma)\xi_o,$$

where $p_j' \in W_{j-1} \cap W_j$. ($\delta_o = \delta_k$ is the trivial path.)

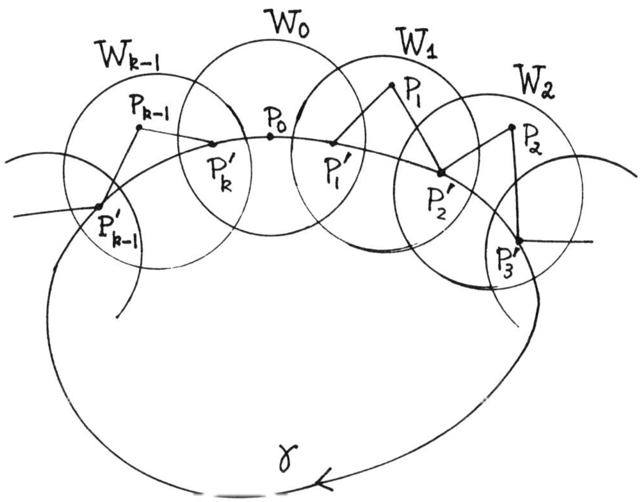

Thus the global function f defined by the analytic continuation of $f = \xi_o$ can be regarded as a holomorphic (resp. meromorphic) function f on $\tilde{M}^*$ which satisfies

$$\varphi^* f = R(\varphi)f \quad \text{for all} \quad \varphi \in G_{\tilde{\pi}}.$$

The correspondence $\xi \longmapsto f$ is (up to constant multiples) nothing but the identification in Lemma 2.2.2.

THEOREM 2.2.3 (Manin-Deligne [8]). Suppose that $\bar{B}$ is normally crossing. Then S_R can be extended to a holomorphic vector bundle $\bar{S}_R$ on $\bar{M}$.

PROOF. Let W be an open set of $\bar{M}$ with a local coordinate system $(w_1,\ldots,w_n)$ such that

$$W = \{(w_1,\ldots,w_n) \mid |w_j| < 1 \quad \text{for} \quad 1 \leq j \leq n\},$$
$$\bar{B} \cap W = \{(w_1,\ldots,w_n) \in W \mid w_\ell \cdots w_n = 0\}.$$

We may assume that, say

$$\bar{D}_j \cap W = \{(w_1,\ldots,w_n) \in W \mid w_j = 0\}$$

for $\ell \leq j \leq n$. Choose a point $p' \in W - \bar{B}$. Let $\hat{\beta}_j$ $(\ell \leq j \leq n)$ be a closed path in $W - \bar{B}$ which starts from p' and rounds $\bar{D}_j$ once in a counterclockwise direction. Let α be a path in $\dot{M}^*$ from p_o to p'. Put

$$\hat{\gamma}_j = \hat{\alpha}\hat{\beta}_j\hat{\alpha}^{-1}.$$

Then $\hat{\gamma}_j$ is a closed path in M^* which starts from p_o. We may regard $\hat{\gamma}_j$ as an element of $G_{\tilde{\pi}}$. Then $\hat{\gamma}_j$ is conjugate to γ_j in section 1.3. Hence $(\hat{\gamma}_j)^{e_j} = 1$. Moreover,

$$\hat{\gamma}_\ell,\ldots,\hat{\gamma}_n$$

commute with each other. Hence, (by the consideration of Jordan canonical forms), there are $(r \times r)$-matrices A_j $(\ell \leq j \leq n)$ such that (i) they commute with each other, (ii) $\exp(2\pi\sqrt{-1}A_j) = R(\hat{\gamma}_j)$ and (iii) the real part $\mathrm{Re}(\lambda)$ of every eigenvalue λ of A_j satisfies $0 \leq \mathrm{Re}(\lambda) < 1$.

Now, if there is a holomorphic vector bundle $\bar{S}_R$ on $\bar{M}$ which extends S_R, then we may assume that there is a local trivialization

$$\begin{array}{ccc} \Psi : \bar{S}_R \mid W & \longrightarrow & W \times \mathbb{C}^r \\ & \searrow \ \circlearrowleft \ \swarrow & \\ & W & \end{array}$$

on W. Ψ induces a local trivialization

$$\begin{array}{ccc} \Psi^* : S_R \mid (W-B) & \longrightarrow & (W-B) \times \mathbb{C}^r \\ & \searrow \ \circlearrowleft \ \swarrow & \\ & W-B & \end{array}$$

on W-B. Ψ^* induces again a local trivialization

$$\begin{array}{ccc} \tilde{\Psi}^* : (\tilde{M}^* \times \mathbb{C}^r) \mid (U-\tilde{\pi}^{-1}(B)) = (U-\tilde{\pi}^{-1}(B)) \times \mathbb{C}^r & \longrightarrow & (U-\tilde{\pi}^{-1}(B)) \times \mathbb{C}^r \\ & \searrow \ \circlearrowleft \ \swarrow & \\ & U-\tilde{\pi}^{-1}(B) & \end{array}$$

where U is a connected component of $\tilde{\pi}^{-1}(W)$. $\tilde{\Psi}^*$ can be written as

$$\tilde{\Psi}^*(p,\xi) = (p,\psi(p)\xi) \quad \text{for} \quad (p,\xi) \in (U-\tilde{\pi}^{-1}(B)) \times \mathbb{C}^r,$$

where

$$\psi : U - \tilde{\pi}^{-1}(B) \longrightarrow GL(r,\mathbb{C})$$

is a holomorphic map. Hence Ψ^* can be written as

$$\Psi^*([p,\xi]) = (\tilde{\pi}(p), \psi(p)\xi),$$

where $[p,\xi] \in S_R$ is the equivalence class of (p,ξ) under the action of $G_{\tilde{\pi}}$. If $[p,\xi] = [q,\eta]$, then $q = \varphi(p)$ and $\eta = R(\varphi)\xi$ for some $\varphi \in G_{\tilde{\pi}}$. Hence

$$(\tilde{\pi}(p),\psi(p)\xi) = \Psi^*([p,\xi]) = \Psi^*([q,\eta]) = (\tilde{\pi}(q),\psi(q)\eta)$$

$$= (\tilde{\pi}(q),\psi(\varphi(p))R(\varphi)\xi).$$

Hence ψ must satisfy the following condition:

$$\psi(p) = \psi(\varphi(p))R(\varphi) \qquad \ldots(1)$$

for $p \in U - \tilde{\pi}^{-1}(B)$ and $\varphi \in G_{\tilde{\pi}}$ such that $\varphi(p) \in U - \tilde{\pi}^{-1}(B)$.

Note that $\tilde{\pi} : U - \tilde{\pi}^{-1}(B) \longrightarrow W-B$ is unramified and $\pi_1(W-B,\ p')$ is a free abelian group generated by $\hat{\beta}_\ell,\ldots,\hat{\beta}_n$. Hence we may assume that an element $\varphi \in G_{\tilde{\pi}}$, such that $\varphi(p) \in U - \tilde{\pi}^{-1}(B)$ for $p \in U - \tilde{\pi}^{-1}(B)$, must be written as

$$\varphi = (\hat{\gamma}_\ell)^{a_\ell}\ldots(\hat{\gamma}_n)^{a_n} \qquad (a_j \in \mathbb{Z} \quad \text{for} \quad \ell \leq j \leq n).$$

Thus the condition (1) is satisfied for ψ if and only if the following condition for ψ is satisfied:

$$\psi(p) = \psi(\hat{\gamma}_j(p))R(\hat{\gamma}_j) \quad \text{for} \quad \ell \leq j \leq n. \qquad \ldots(2)$$

Next, let W' with $W' \subset W$ be an open set of $\bar{M}$ with a coordinate system $(w_1',\ldots,w_n')$ such that

$$W' - \{(w_1',\ldots,w_n') \mid |w_j'| < 1 \quad \text{for} \quad 1 \leq j \leq n\},$$

$$\bar{B} \cap W' - \{(w_1',\ldots,w_n') \in W' \mid w_m'\ldots w_n' - 0\},$$

$(k \leq m \leq n)$. Let

$$\begin{array}{ccc} \Psi' : \bar{S}_R|W' & \longrightarrow & W' \times \mathbb{C}^r \\ & \searrow \circlearrowleft \swarrow & \\ & W' & \end{array}$$

be a local trivialization with respect to W'. Then we may write

$$\Psi.\Psi'^{-1}(p,\xi) = (p,g_{WW'}(p)\xi) \quad \text{for} \quad (p,\xi) \in W' \times \mathbb{C}^r,$$

where

$$g_{WW'} : W' \longrightarrow GL(r,\mathbb{C})$$

is a holomorphic map, which must clearly satisfy the following compatibility condition:

$$\psi(p) = g_{WW'}(\tilde{\pi}(p))\psi'(p) \quad \text{for} \quad p \in U' - \tilde{\pi}^{-1}(B) \qquad \ldots(3)$$

where U' is a connected component of $\tilde{\pi}^{-1}(W')$ such that $U' \subset U$ and $\psi' : U' - \tilde{\pi}^{-1}(B) \longrightarrow GL(r,\mathbb{C})$ is the holomorphic map defined as ψ with respect to W'.

Note that the set $\{g_{jk}, \psi, g_{WW'}\}$ is a set of transition matrices for $\bar{S}_R$. ($\{g_{jk}\}$ is the set of transition matrices for S_R defined before.) More precisely, ψ induces the transition matrix g_{WW_j} on $W \cap W_j$.

Conversely, if there are holomorphic maps

$$\psi : U - \tilde{\pi}^{-1}(B) \longrightarrow GL(r,\mathbb{C})$$

and

$$g_{WW'} : W' \longrightarrow GL(r,\mathbb{C}),$$

which satisfy the conditions (2) and (3), then the set $\{g_{jk}, \psi, g_{WW'}\}$ is a set of transition matrices of a holomorphic vector bundle $\bar{S}_R$ on $\bar{M}$ which is an extension of S_R.

We now find such ψ and $g_{WW'}$. We may write

$$w_j = w'_j\, g_j(w'_1,\ldots,w'_n) \quad \text{for} \quad m \leq j \leq n,$$

$$w_j = g_j(w'_1,\ldots,w'_n) \quad \text{for} \quad \ell \leq j \leq m-1,$$

on W', where g_j $(\ell \leq j \leq n)$ are non-vanishing holomorphic functions on W'. Let $h_j(w'_1,\ldots,w'_n)$ $(\ell \leq j \leq n)$ be holomorphic functions on W' such that

$$g_j = \exp(h_j) \quad \text{for} \quad \ell \leq j \leq n.$$

Note that, since $U' \subset U$,

$$\hat{\gamma}'_j = \hat{\gamma}_j \in G_{\tilde{\pi}} \quad \text{for} \quad m \leq j \leq n,$$

where $\hat{\gamma}'_j$ is an element of $G_{\tilde{\pi}}$ defined as $\hat{\gamma}_j$ with respect to W'. We put

$$\psi = \prod_{j=\ell}^{n} w_j^{-A_j} = \exp(-\sum_{j=\ell}^{n} A_j \log(w_j)),$$

$$g_{WW'} = \exp(-\sum_{j=\ell}^{n} A_j h_j).$$

Then

$$\hat{\gamma}_j\psi = \hat{\gamma}_j^*(-\sum_{j=\ell}^{n} A_j \log(w_j)))$$

$$= \exp(-A_\ell \log(w_\ell)-\ldots-(A_j \log(w_j)+A_j 2\pi\sqrt{-1})-\ldots-A_n \log(w_n))$$

$$= \psi \exp(-A_j 2\pi\sqrt{-1}) = \psi R(\hat{\gamma}_j)^{-1} \quad \text{for} \quad \ell \leq j \leq n.$$

$$\psi = \prod_{j=\ell}^{m-1} w_j^{-A_j} \prod_{j=m}^{n} w_j^{-A_j} = \exp(-\sum_{j=\ell}^{m-1} A_j h_j)\exp(-\sum_{j=m}^{n} A_j h_j) \prod_{j=m}^{n} w'_j{}^{-A_j}$$

$$= g_{WW'}\, \psi'.$$

Thus our ψ's and $g_{WW'}$'s satisfy the conditions (2) and (3). □

As is easily seen, the above proof works for any representation

$$R : \pi_1(\bar{M}-\bar{B},\ p_o) \longrightarrow GL(r,\mathbb{C})$$

and the flat vector bundle S_R on $\bar{M}-\bar{B}$ induced by R. In particular, the conclusion holds even if Condition 1.2.4 on (M,D) does not hold. We restate Theorem 2.2.3 in this general form:

THEOREM 2.2.3' Suppose that $\bar{B}$ is normally crossing. Then, for any representation R of $\pi_1(\bar{M}-\bar{B},\ p_o)$, the flat vector bundle S_R induced by R can be extended to a holomorphic vector bundle $\bar{S}_R$ on $\bar{M}$.

For a general case in which $\bar{B}$ may not be normally crossing, there may not exist a holomorphic vector bundle (that is, a locally free sheaf) on $\bar{M}$ which is an extension of S_R (= $\mathcal{O}(S_R)$, identified). But

THEOREM 2.2.4 $\mathcal{O}(S_R)$ can be extended to a coherent analytic sheaf S on $\bar{M}$ such that $S|M$ is locally free.

PROOF. By Hironaka's theorem, there is a proper bimeromorphic holomorphic map

$$\tau : \hat{M} \longrightarrow \bar{M}$$

of a projective manifold $\hat{M}$ onto $\bar{M}$ such that (ii) $\tau : \hat{M}-\tau^{-1}(\mathrm{Sing}\bar{B}) \longrightarrow \bar{M} - \mathrm{Sing}\bar{B}$ is biholomorphic and (ii) $\hat{B} = \tau^{-1}(\bar{B})$ is a normally crossing hypersurface of $\hat{M}$. By Theorem 2.2.3', there is a locally free sheaf $\mathcal{O}(\hat{S}_R)$ on $\hat{M}$ which extends $\tau^*(\mathcal{O}(S_R))$ ($\simeq \mathcal{O}(S_R)$) on $\hat{M} - \hat{B}$. By Grauert's theorem (Grauert [19]), $S = \tau_*\mathcal{O}(\hat{S}_R)$ is a coherent sheaf on $\bar{M}$, which extends $\tau_*\tau^*\mathcal{O}(S_R) = \mathcal{O}(S_R)$ and is locally free on M by (i). □

For meromorphic sections $\xi_1,\ldots,\xi_r$ on M of $S|M$, let $\det(\xi_1\ldots\xi_r)$ be the meromorphic section on M of the line bundle $\Lambda^r(S|M)$ defined by

$$\det(\xi_1\ldots\xi_r) = \{\det(\xi_{1j}\ldots\xi_{rj})\},$$

where $\xi_\nu = \{\xi_{\nu j}\}$ for $1 \leq \nu \leq r$.

THEOREM 2.2.5 (cf. Nakano [43]) For the coherent sheaf S on $\bar{M}$ in Theorem 2.2.4 and for any point $p \in M$, there are meromorphic sections $\xi_1,\ldots,\xi_r$ on $\bar{M}$ of S such that $\xi_1,\ldots,\xi_r$ are holomorphic around p and $\det((\xi_1|M)\ldots(\xi_r|M))(p) \neq 0$.

PROOF. Let

$$\tau : Q_p(\bar{M}) \longrightarrow \bar{M}$$

be the blowing up of $\bar{M}$ at p. Put $E = \tau^{-1}(p)$, the exceptional hypersurface. Let H be a hyperplane bundle on $\bar{M}$ ($\subset \mathbb{P}^N$). Then, since τ^*H is ample, there is a positive integer m such that

$$H^1(Q_p(\bar{M}), \mathcal{O}([E]^{-1}) \otimes \tau^*S \otimes (\tau^*\mathcal{O}(H))^m) = 0. \qquad \ldots(1)$$

Note that

$$(\tau^*S \otimes (\tau^*\mathcal{O}(H))^m)|E = \tau^*(S \otimes \mathcal{O}(H^m))|E$$

is isomorphic to the trivial bundle on E of rank r. Hence, by (1) and by the exact sequence

$$0 \longrightarrow \mathcal{O}([E]^{-1}) \otimes \tau^*(S \otimes \mathcal{O}(H^m)) \longrightarrow \tau^*(S \otimes \mathcal{O}(H^m)) \longrightarrow \mathcal{O}_E^r \longrightarrow 0,$$

we have the exact sequence

$$H^0(Q_p(\bar{M}),\tau^*(S \otimes \mathcal{O}(H^m))) \longrightarrow H^0(E,\mathcal{O}_E^r) \longrightarrow 0. \qquad \ldots(2)$$

Let $\{\hat{\eta}_1,\ldots,\hat{\eta}_d\}$ be a basis of $H^0(Q_p(\bar{M}),\tau^*(S \otimes \mathcal{O}(H^m)))$. (2) means that $\hat{\eta}_1|E,\ldots,\hat{\eta}_d|E$ can be regarded as constant vectors such that the matrix $(\hat{\eta}_1|E\ldots\hat{\eta}_d|E)$ has the rank r. Hence $d \geq r$ and we may assume that

$$\det(\hat{\eta}_1|E\ldots\hat{\eta}_r|E) \neq 0.$$

Since $\hat{\eta}_\nu|E$ can be regarded as constant vectors, if we define η_ν by

$$\eta_\nu = \begin{cases} \hat{\eta}_\nu & \text{on } \bar{M} - p \quad (\simeq Q_p(\bar{M}) - E), \\ \hat{\eta}_\nu|E & \text{at } p, \end{cases}$$

then η_ν $(1 \leq \nu \leq r)$ are holomorphic sections on $\bar{M}$ of $S \otimes \mathcal{O}(H^m)$ such that

$$\det((\eta_1|M)\ldots(\eta_r|M))(p) \neq 0.$$

Now, let ζ be a holomorphic section of H such that $\zeta(p) \neq 0$. Then

$$\xi_1 = \eta_1/\zeta^m, \ldots, \xi_r = \eta_r/\zeta^m$$

are meromorphic sections on $\bar{M}$ of S, which satisfy the requirement. □

REMARK 2.2.6 For given points $p_1, \ldots, p_t$ on M, if we consider the blowing up

$$\tau : Q(\bar{M}) \longrightarrow \bar{M}$$

of $\bar{M}$ at $p_1, \ldots, p_t$, then a similar proof to the proof of Theorem 2.2.5 shows that there are meromorphic sections $\xi_1, \ldots, \xi_r$ of S such that they are holomorphic around every p_j and

$$\det((\xi_1|M)\ldots(\xi_r|M))(p_j) \neq 0 \quad \text{for} \quad 1 \leq j \leq t.$$

Now we are ready to prove Theorem 2.2.1.

PROOF OF THEOREM 2.2.1 By Theorem 2.2.5, there are meromorphic sections $\xi_1, \ldots, \xi_r$ on $\bar{M}$ of S such that they are holomorphic around p_o and

$$\det((\xi_1|M)\ldots(\xi_r|M))(p_o) \neq 0.$$

As before, we cover $M^* = M-B$ by open sets W_j such that $S_R = S|M^*$ is trivial on W_j. ($p_o \in W_o$.) We write

$$\xi_\nu|M^* = \xi_\nu^* = \{\xi_{\nu j}\}$$

and put

$$f_\nu = \xi_{\nu o} \quad \text{for} \quad 1 \leq \nu \leq r.$$

By the analytic continuation of f_ν as before, we have r-vector valued meromorphic functions f_ν on $\tilde{M}^*$ such that

$$\varphi^* f_\nu = R(\varphi) f_\nu \quad \text{for all } \varphi \in G_{\tilde{\pi}}.$$

We show that every f_ν can be continued meromorphically on $\tilde{M}$. Take a point $p \in B$. Suppose that $p \in D_n$, say. Let W be an open neighbourhood of p in M such that S is trivial on W. Let $(w_1, \ldots, w_n)$ and $(z_1, \ldots, z_n)$ be coordinate systems on W and U, a connected component of $\pi^{-1}(W)$, respectively, such that $\tilde{\pi}$ is locally given by

$$\tilde{\pi} : (z_1, \ldots, z_n) \in U \longmapsto (w_1, \ldots, w_n) = (z_1, \ldots, z_{n-1}, z_n^{e_n}) \in W.$$

Let

$$\psi = w_n^{-A_n} = \exp(-A_n \log(w_n))$$

be the holomorphic map of

$$U - \tilde{\pi}^{-1}(B) = \{(z_1,\dots,z_n) \in U \mid z_n \neq 0\}$$

into $GL(r,\mathbb{C})$, which is defined in the proof of Theorem 2.2.3. Since

$$R(\hat{\gamma}_n)^{e_n} = R(\hat{\gamma}_n^{e_n}) = R(1) = 1,$$

the matrix $R(\hat{\gamma}_n) = \exp(2\pi\sqrt{-1}A_n)$ is semi-simple. Hence A_n can be chosen so that there is a non-singular $(r \times r)$-matrix C such that

$$CA_nC^{-1} = \begin{pmatrix} \lambda_1 & & 0 \\ & \ddots & \\ 0 & & \lambda_r \end{pmatrix} \quad \text{and} \quad CR(\hat{\gamma}_n)C^{-1} = \begin{pmatrix} \exp(2\pi\lambda_1\sqrt{-1}) & & 0 \\ & \ddots & \\ 0 & & \exp(2\pi\lambda_r\sqrt{-1}) \end{pmatrix}.$$

Here $\lambda_1,\dots,\lambda_r$ are eigenvalues of A_n such that

$$0 \leq \operatorname{Re}(\lambda_\nu) < 1 \quad \text{for} \quad 1 \leq \nu \leq r.$$

Since

$$\exp(2\pi\sqrt{-1}e_n\lambda_\nu) = 1 \quad \text{for} \quad 1 \leq \nu \leq r,$$

every $e_n\lambda_\nu$ must be an integer. We may write

$$\lambda_\nu = a_\nu/e_n \quad \text{for} \quad 1 \leq \nu \leq r,$$

where a_ν are integers such that

$$0 \leq a_\nu \leq e_n-1.$$

Now,

$$\psi = \exp(-C^{-1}(CA_nC^{-1})C\log(w_n))$$

$$= C^{-1} \begin{pmatrix} \exp((-a_1/e_n)\log(w_n)) & & 0 \\ & \ddots & \\ 0 & & \exp((-a_r/e_n)\log(w_n)) \end{pmatrix} C$$

$$= C^{-1} \begin{pmatrix} z_n^{-a_1} & & 0 \\ & \ddots & \\ 0 & & z_n^{-a_r} \end{pmatrix} C \qquad \text{on } U - \tilde{\pi}^{-1}(B).$$

We extend ψ meromorphically on U by putting

$$\psi = C^{-1}\begin{pmatrix} z_n^{-a_1} & & 0 \\ & \ddots & \\ 0 & & z_n^{-a_r} \end{pmatrix} C \qquad \text{on } U.$$

Then

$$\psi^{-1} = C^{-1}\begin{pmatrix} z_n^{a_1} & & 0 \\ & \ddots & \\ 0 & & z_n^{a_r} \end{pmatrix} C$$

is holomorphic on U.

Now, let γ be a path from p_0 to p, which is contained in M* except the end point p. Suppose that $\{W_0, W_1, \ldots, W_t, W\}$ covers γ.

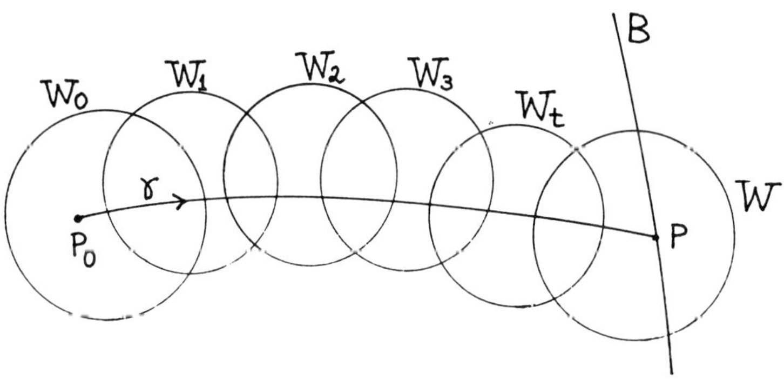

Using the previous notations, put

$$f_\nu(q) = g_{01}g_{12}\cdots g_{t-1,t}\psi^{-1}(q)\xi_{\nu W}(\tilde{\pi}(q)) \quad \text{for} \quad q \in U,$$

where $\xi_{\nu W}$ is the representative of ξ_ν with respect to the trivialization of S on W. Then this is clearly the meromorphic continuation on U of f_ν on $\tilde{M}$*.

Thus we have the representative vector functions

$$f_1, \ldots, f_r$$

with respect to R such that they are holomorphic around every point $\tilde{p}_0$ in $\tilde{\pi}^{-1}(p_0)$ and

$$\det(f_1(\tilde{p}_0)\ldots f_r(\tilde{p}_0)) \neq 0.$$

Hence

$$F = (f_1 \dots f_r)$$

is a non-degenerate representative matrix function with respect to R. □

REMARK 2.2.7 (1) The idea of the above proof of Theorem 2.2.1 comes from Kita [39]. Theorem 2.2.1 can, in fact, be regarded as a special case of the solution of the *Riemann-Hilbert problem* (cf. also Röhrl [52]).
(2) if $\tilde{M}$ is biholomorphic to a bounded domain in $\mathbb{C}^n$, then Theorem 2.2.1 can be proved by using a *theta-Fuchs series* and a *zeta-Fuchs series of Poincaré*, (see Siegel [57]).

Now, we take a non-degenerate representative matrix valued meromorphic function F with respect to R and fix it.

LEMMA 2.2.8 If F_1 is another non-degenerate representative matrix function, then there is an $(r \times r)$-matrix valued function P on M such that $\det(P)$ is not identically zero and $F_1 = FP$. Conversely, for any $(r \times r)$-matrix valued meromorphic function P on M such that $\det(P)$ is not identically zero, FP is a non-degenerate representative matrix function.

PROOF. Put $P = F^{-1}F_1$. Then P is an $(r \times r)$-matrix valued meromorphic function on $\tilde{M}$. For any $\varphi \in G_{\tilde{\pi}}$,

$$\varphi^*P = \varphi^*(F^{-1}F_1) = (\varphi^*F)^{-1}(\varphi^*F_1) = F^{-1}R(\varphi)^{-1}R(\varphi)F_1 = F^{-1}F_1 = P.$$

Hence P is (regarded as) a matrix valued meromorphic function on M such that $\det(P)$ is not identically zero.

The converse is clear. □

We consider the extension field

$$K_R = \mathbb{C}(M)(f_{11}, f_{12}, \dots, f_{rr})$$

of $\mathbb{C}(M)$, where $F = (f_{jk})$. Then, by Lemma 2.2.8, K_R does not depend on the choice of a representative matrix function F, but depends only on R. Moreover, we have clearly

$$\varphi^*(K_R) = K_R \quad \text{for all} \quad \varphi \in G_{\tilde{\pi}}.$$

It should be noted that $K_R/\mathbb{C}(M)$ may not in general be a finite extension.

PROPOSITION 2.2.9 $G(K_R) = \ker(R)$, where

$$G(K_R) = \{\varphi \in G_{\tilde{\pi}} \mid \varphi^*f = f \quad \text{for all } f \in K_R\}.$$

Moreover, if the image $\mathrm{Im}(R)$ of R is a finite subgroup of $GL(r,\mathbb{C})$, then $K_R = K(\ker(R))$, where

$$K(\ker(R)) = \{f \in \mathbb{C}(\tilde{M}) \mid \varphi^*f = f \quad \text{for all } \varphi \in \ker(R)\}.$$

In this case, $K_R/\mathbb{C}(M)$ is a finite Galois extension whose Galois group $\mathrm{Gal}(K_R/\mathbb{C}(M))$ is isomorphic to $\mathrm{Im}(R)$.

PROOF. If φ is an element of $\ker(R)$, then

$$\varphi^*F = R(\varphi)F = F.$$

Hence

$$\varphi^*f_{jk} = f_{jk} \quad \text{for} \quad 1 \leq j,k \leq r.$$

This implies that

$$\varphi \in G(K_R).$$

Conversely, if $\varphi \in G(K_R)$, then

$$\varphi^*f_{jk} = f_{jk} \quad \text{for} \quad 1 \leq j,k \leq r.$$

Hence

$$\varphi^*F = F.$$

Then we have

$$1 = (\varphi^*F)F^{-1} = R(\varphi).$$

Hence

$$\varphi \in \ker(R).$$

Thus we have

$$G(K_R) = \ker(R).$$

Now, from this relation, we have

$$K_R \subset K(G(K_R)) = K(\ker(R)).$$

Suppose that $\mathrm{Im}(R) \simeq G_{\tilde{\pi}}/\ker(R)$ is a finite group. The, by Theorem 2.1.3, $K(\ker(R))/\mathbb{C}(M)$ is a finite Galois extension. Hence $K_R/\mathbb{C}(M)$ is a finite extension. By Theorem 2.1.3 again,

$$K_R = K(G(K_R)) = K(\ker(R))$$

and

$$\mathrm{Gal}(K_R/\mathbb{C}(M)) \simeq G_{\tilde{\pi}}/\ker(R) \simeq \mathrm{Im}(R). \quad \square$$

Conversely

THEOREM 2.2.10 Every finite Galois extension K of $\mathbb{C}(M)$ with $K \subset \mathbb{C}(\tilde{M})$ can be written as $K = K_R$ for some representation R of $G_{\tilde{\pi}}$ such that $\ker(R) = G(K)$.

PROOF. By Theorem 2.1.3, $G(K)$ is a normal subgroup of $G_{\tilde{\pi}}$ of finite index such that

$$\mathrm{Gal}(K/\mathbb{C}(M)) \simeq G_{\tilde{\pi}}/G(K).$$

Let r be a positive integer and

$$R_1 : \mathrm{Gal}(K/\mathbb{C}(M)) \longrightarrow GL(r,\mathbb{C})$$

be a *faithful* representation, that is, a representation such that

$$\ker(R_1) = \{1\}.$$

(Such an R_1 exists. For example, put $r = [K : \mathbb{C}(M)]$ and let

$$R_1 : \mathrm{Gal}(K/\mathbb{C}(M)) \longrightarrow GL(r,\mathbb{C})$$

be the regular representation. Then R_1 is faithful.)

Then the composition

$$R : G_{\tilde{\pi}} \longrightarrow G_{\tilde{\pi}}/G(K) \simeq \mathrm{Gal}(K/\mathbb{C}(M)) \xrightarrow{R_1} GL(r,\mathbb{C})$$

of homomorphisms is a representation such that

$$\ker(R) = G(K).$$

Hence, by Proposition 2.2.9 and Theorem 2.1.3,

$$K_R = K(\ker(R)) = K(G(K)) = K. \quad \square$$

This theorem means that any finite Galois extension K of $\mathbb{C}(M)$ with $K \subset \mathbb{C}(\tilde{M})$ can be obtained by attaching some representative functions to $\mathbb{C}(M)$.

LEMMA 2.2.11 Let R and R' be equivalent representations of $G_{\tilde{\pi}}$, that is, there is a non-singular $(r \times r)$-matrix C such that

$$R'(\varphi) = CR(\varphi)C^{-1} \quad \text{for all} \quad \varphi \in G_{\tilde{\pi}} .$$

If F is a non-degenerate representative matrix function with respect to R, then CF is one with respect to R'. In particular, $K_{R'} = K_R$.

PROOF. $\varphi^*(CF) = C\varphi^*F = CR(\varphi)F = (CR(\varphi)C^{-1})(CF)$

$$= R'(\varphi)(CF). \quad \text{for all } \varphi \in G_{\tilde{\pi}} . \quad \square$$

PROPOSITION 2.2.12 Any finite Galois extension K of $\mathbb{C}(M)$ with $K \subset \mathbb{C}(\tilde{M})$ can be obtained by attaching to $\mathbb{C}(M)$ representative functions with respect to some irreducible unitary representations:

$$R : G_{\tilde{\pi}} \longrightarrow \mathrm{Gal}(K/\mathbb{C}(M)) \longrightarrow U(r) \subset GL(r,\mathbb{C}).$$

PROOF A (faithful) representation R_1 of the finite group $\mathrm{Gal}(K/\mathbb{C}(M))$ is equivalent to a unitary representation, which is completely reducible. Hence, by Theorem 2.2.10 and Lemma 2.2.11, we obtain the assertion. $\square$

For the rest of this section, we consider the case:

$\bar{B} = \emptyset$ (empty),

$M = \bar{M}$ is a projective manifold and

$\tilde{\pi} = \tilde{M} \longrightarrow M$ is the universal covering of M.

As an application of Theorem 2.2.1, we give a proof of the famous *algebraic de Rham theorem*. First of all,

DEFINITION 2.2.13 A meromorphic 1-form η on a projective manifold M is called an *abelian differential of the second kind* if (1) $d\eta = 0$ and (2) for any non-singular point p of $\mathrm{Supp}(D_\infty(\eta))$, the support of the polar divisor $D_\infty(\eta)$ and for any closed small path γ in $M - \mathrm{Supp}(D_\infty(\eta))$ which rounds $\mathrm{Supp}(D_\infty(\eta))$ around p once in a counterclockwise direction,

$$\int_\gamma \eta = 0$$

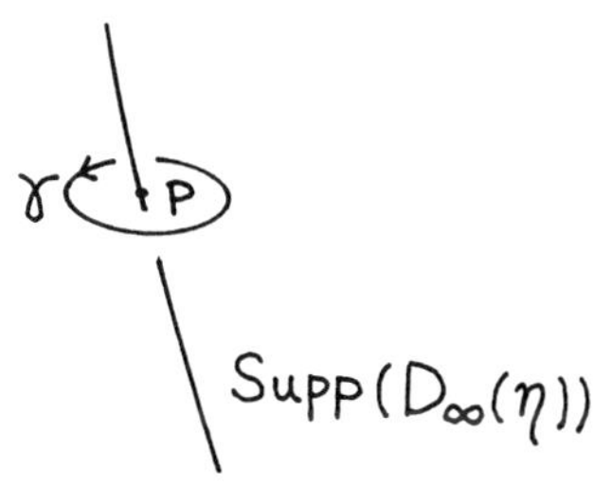

(In other words, η has no residue along $\mathrm{Supp}(D_\infty(\eta))$).

The set $A_2(M)$ of all abelian differentials of the second kind forms a complex vector space in a natural way. If f is a meromorphic function on

M, then the total differential df of f is an abelian differential of the second kind.

$$\mathcal{B}(M) = \{df \mid f \in \mathbb{C}(M)\}$$

is a complex vector subspace of $A_2(M)$.

THEOREM 2.2.14 (Algebraic de Rham Theorem) For a projective manifold M, put $A_2(M) = \mathcal{A}_2(M)/\mathcal{B}(M)$. Then the bilinear pairing

$$(\eta,\gamma) \in A_2(M) \times H_1(M;\mathbb{C}) \longmapsto \int_\gamma \eta \in \mathbb{C}$$

is well defined and non-degenerate. Moreover, $A_2(M)$ is isomorphic to the dual space $H_1(M;\mathbb{C})^*$ $(\simeq H^1(M;\mathbb{C}))$ of $H_1(M;\mathbb{C})$ by this pairing.

For the proof of the theorem, we first give

DEFINITION 2.2.15 Let $\tilde{\pi} : \tilde{M} \longrightarrow M$ be the universal covering of a projective manifold M. A meromorphic function f on $\tilde{M}$ is called an *additive function* if

$$\varphi^* f = f + c_f(\varphi) \quad \text{for all} \quad \varphi \in G_{\tilde{\pi}} (\simeq \pi_1(M,p_o)),$$

where $c_f(\varphi)$ is a complex number.

It is clear that $c_f(\varphi)$ must satisfy the following condition:

$$c_f(\varphi_1\varphi_2) = c_f(\varphi_1) + c_f(\varphi_2) \quad \text{for} \quad \varphi_1,\varphi_2 \in G_{\tilde{\pi}}.$$

Hence c_f is a homomorphism of $G_{\tilde{\pi}}$ into $\mathbb{C}$. Since $\mathbb{C}$ is abelian, this homomorphism splits as follows:

$$\begin{array}{ccc} G_{\tilde{\pi}} \simeq \pi_1(M, p_o) & \longrightarrow & \pi_1(M, p_o)/[\pi_1,\pi_1] \\ & {\scriptstyle c_f} \searrow \quad \swarrow {\scriptstyle \hat{c}_f} & \\ & \mathbb{C} & \end{array}$$

where $[\pi_1,\pi_1]$ is the commutator group of $\pi_1 = \pi_1(M, p_o)$. It is well known that $\pi_1(M, p_o)/[\pi_1,\pi_1]$ is naturally isomorphic to $H_1(M;\mathbb{Z})$. Note that if $\gamma \in H_1(M;\mathbb{Z})$ is a torsion, then $\hat{c}_f(\gamma) = 0$. Hence, by the coefficient extension, we have the homomorphism

$$\hat{c}_f : H_1(M;\mathbb{C}) \longrightarrow \mathbb{C}.$$

If f is an additive function, then

$$\varphi^*(df) = df \quad \text{for all} \quad \varphi \in G_{\tilde{\pi}}.$$

Hence df is (regarded as) a meromorphic 1-form on M, which is clearly closed. Locally, f can be regarded as a meromorphic function on open sets of M. Hence df has no residue along $\mathrm{Supp}(D_\infty(df))$. Thus df is an abelian differential of the second kind. Moreover, we have easily

$$c_f(\gamma) = \int_\gamma df$$

and

$$f(q) = f(q_0) + \int_\delta df,$$

where γ (resp. δ) is a closed path (resp. a path which corresponds to $q \in \tilde{M}$) which starts from p_0. (q_0 = the homotopy class of the trivial path $\widehat{p_0 p_0}$.) Here γ and δ are chosen so that they do not pass through $\mathrm{Supp}(D_\infty(df))$.

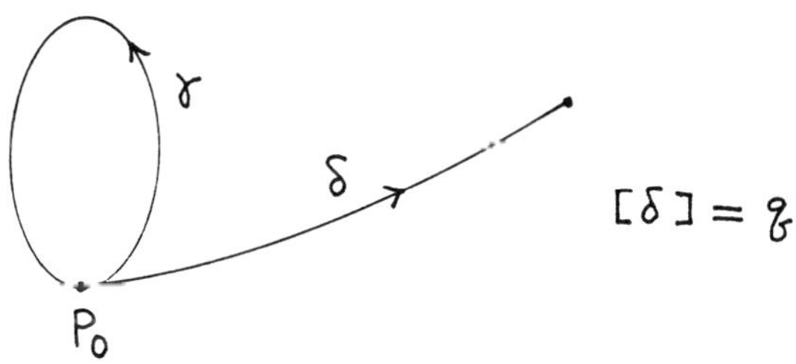

Conversely, let η be an abelian differential of the second kind. Put

$$f(q) = \int_\delta \eta \quad \text{for} \quad q \in \tilde{M} - \tilde{\pi}^{-1}(\mathrm{Supp}(D_\infty(\eta))),$$

where δ is a path which starts from p_0, corresponds to $q \in \tilde{M}$ and does not pass through $\mathrm{Supp}(D_\infty(\eta))$. Then f is a holomorphic function. Since η is meromorphic and has no residue along $\mathrm{Supp}(D_\eta(\eta))$, f can be extended to a meromorphic function f on $\tilde{M}$ in a natural way. f is then an additive function. In fact, for a closed path γ which strats from p_0 and does not pass through $\mathrm{Supp}(D_\infty(\eta))$,

$$(\gamma^*f)(q) = f(\gamma(q)) = \int_\gamma \eta + \int_\delta \eta = \int_\gamma \eta + f(q).$$

Thus the pairing in Theorem 2.2.14 is nothing but the one induced by the bilinear pairing

$$(\eta,\gamma) \in A_2(M) \times H_1(M,\mathbb{C}) \longmapsto \hat{c}_f(\gamma) \in \mathbb{C},$$

where $df = \eta$.

Suppose that $\hat{c}_f(\gamma) = 0$ for all $\gamma \in H_1(M,\mathbb{C})$. Then

$$\varphi^* f = f \quad \text{for all} \quad \varphi \in G_{\tilde{\pi}}.$$

Hence f is a meromorphic function on M. Hence $\eta = df \in \mathcal{B}(M)$.

In order to prove the non-degeneracy of the pairing and the last assertion of Theorem 2.2.14, it clearly suffices to show the following lemma.

<u>LEMMA 2.2.16</u> Let $\gamma_1,\ldots,\gamma_{b_1}$ be a free basis of $H_1(M;\mathbb{Z})$. (b_1 = the first Betti number of M.) Then, for any complex numbers, $c_1,\ldots,c_{b_1}$, there is an abelian differential η of the second kind such that

$$\int_{\gamma_j} \eta = c_j \quad \text{for} \quad 1 \leq j \leq b_1.$$

<u>PROOF</u>. Let

$$R_1 : H_1(M;\mathbb{Z}) \longrightarrow GL(2,\mathbb{C})$$

be the homomorphism defined by

$$R_1(\gamma_j) = \begin{pmatrix} 1 & c_j \\ 0 & 1 \end{pmatrix}$$

and

$$R_1(\gamma) = 0 \quad \text{for all torsion } \gamma.$$

Let

$$R : G_{\tilde{\pi}} \simeq \pi_1(M, p_o) \longrightarrow H_1(M;\mathbb{Z}) \xrightarrow{R_1} GL(2,\mathbb{C})$$

be the composition of the homomorphisms. Let

$$F = \begin{pmatrix} f_{11} & f_{12} \\ f_{21} & f_{22} \end{pmatrix}$$

be a non-degenerate representative matrix function with respect to R. Then

$$\begin{pmatrix} \gamma_j^* f_{11} & \gamma_j^* f_{12} \\ \gamma_j^* f_{21} & \gamma_j^* f_{22} \end{pmatrix} = \begin{pmatrix} 1 & c_j \\ 0 & 1 \end{pmatrix} \begin{pmatrix} f_{11} & f_{12} \\ f_{21} & f_{22} \end{pmatrix} = \begin{pmatrix} f_{11}+c_j f_{21} & f_{12}+c_j f_{22} \\ f_{21} & f_{22} \end{pmatrix}$$

Hence

$$\gamma_j^*(f_{11}/f_{21}) = f_{11}/f_{21} + c_j$$

and

$$\gamma_j^*(f_{12}/f_{22}) = f_{12}/f_{22} + c_j.$$

Since $f_{11}f_{22} - f_{21}f_{12}$ is not identical to zero, at least one of f_{21} and f_{22} is not identical to zero. Suppose that f_{21}, say, is not identical to zero. Putting $f = f_{11}/f_{21}$, we have

$$\gamma_j^*(f) = f + c_j \quad \text{for} \quad 1 \leq j \leq b_1.$$

Hence f is an additive function on M such that

$$\int_{\gamma_j} (df) = c_j \quad \text{for} \quad 1 \leq j \leq b_1.$$

Thus $\eta = df$ satisfies the condition. □

See Atiyah-Hodge [3] for another proof of Theorem 2.2.14.

Finally, by making use of Lemma 2.2.16, we can give an example of fields L with $\mathbb{C}(M) \subset L \subset \mathbb{C}(\tilde{M})$ such that $[L : \mathbb{C}(M)] = +\infty$ and $K(G(L)) \neq L$, (cf. (2) of Theorem 2.1.3). First of all

LEMMA 2.2.17 Let f be an additive function with non-trivial c_f on a projective manifold M. Then $\mathbb{C}(M)(f)/\mathbb{C}(M)$ is a purely transcendental extension.

PROOF. Suppose that there is an irreducible polynomial $x^m + a_1x^{m-1} + \dots + a_m$ with $a_j \in \mathbb{C}(M)$ for $1 \leq j \leq m$ such that

$$f^m + a_1f^{m-1} + \dots + a_m = 0. \qquad \dots(1)$$

Take an element $\varphi \in G_{\tilde{\pi}}$ (= $\pi_1(M, p_o)$, identified) such that $c_f(\varphi) \neq 0$. Then we have

$$\begin{aligned} 0 &= \varphi^*(f^m + a_1 f^{m-1} + \ldots + a_m) \\ &= (\varphi^* f)^m + a_1(\varphi^* f)^{m-1} + \ldots + a_m \\ &= (f + c_f(\varphi))^m + a_1(f + c_f(\varphi))^{m-1} + \ldots + a_m. \end{aligned}$$

That is

$$f^m + (mc_f(\varphi) + a_1)f^{m-1} + \ldots + (c_f(\varphi)^m + \ldots + a_{m-1}c_f(\varphi) + a_m) = 0 \qquad \ldots(2)$$

By subtracting (2) from (1),

$$mc_f(\varphi)f^{m-1} + \ldots + (c_f(\varphi)^m + \ldots + a_{m-1}c_f(\varphi)) = 0.$$

This contradicts to the irreducibility of the polynomial $x^m + a_1 x^{m-1} + \ldots + a_m$.□

Now, suppose that the first Betti number b_1 of M is positive. Let $c_1, \ldots, c_{b_1}$ be complex numbers which are linearly independent over the field $\mathbb{Q}$ of all rational numbers. By Lemma 2.2.16, there is an additive function f such that

$$\hat{c}_f(\gamma_j) = c_j \quad \text{for} \quad 1 \leq j \leq b_1.$$

Then it is clear that, for $\gamma \in H_1(M;\mathbb{Z})$, $\hat{c}_f(\gamma) = 0$ if and only if γ is a torsion. Put

$$L = \mathbb{C}(M)(f).$$

By Lemma 2.2.17, $[L : \mathbb{C}(M)] = +\infty$. We show that

$$K(G(L)) \neq L.$$

Note first that

$$\begin{aligned} G(L) &= \{\varphi \in \pi_1(M, p_o) \mid \varphi^* f = f\} \\ &= \{\varphi \in \pi_1(M, p_o) \mid c_f(\varphi) = 0\} \\ &= \alpha^{-1}(T), \end{aligned}$$

where T is the torsion subgroup of $H_1(M;\mathbb{Z})$ and

$$\alpha : \pi_1(M, p_o) \longrightarrow H_1(M;\mathbb{Z})$$

is the canonical homomorphism. Note next that $\pi_1(M, p_o)/\alpha^{-1}(T) \simeq H_1(M;\mathbb{Z})/T$ is a free abelian group of rank b_1. Let H be a subgroup of $H_1(M;\mathbb{Z})$ of finite index such that

$$T \subsetneq H \subsetneq H_1(M;\mathbb{Z}).$$

Then

$$[\pi_1,\pi_1] \subset \alpha^{-1}(T) \subsetneq \alpha^{-1}(H) \subsetneq \pi_1(M, p_o)$$

and

$$K(G(L)) = K(\alpha^{-1}(T)) \supset K(\alpha^{-1}(H)) \supsetneq \mathbb{C}(M).$$

$K(\alpha^{-1}(H))/\mathbb{C}(M)$ is then a finite abelian extension. On the other hand, by Lemma 2.2.17 again, every field K such that

$$\mathbb{C}(M) \subsetneq K \subset L = \mathbb{C}(M)(f)$$

is purely transcendental. Hence

$$K(G(L)) = K(\alpha^{-1}(T)) \neq L.$$

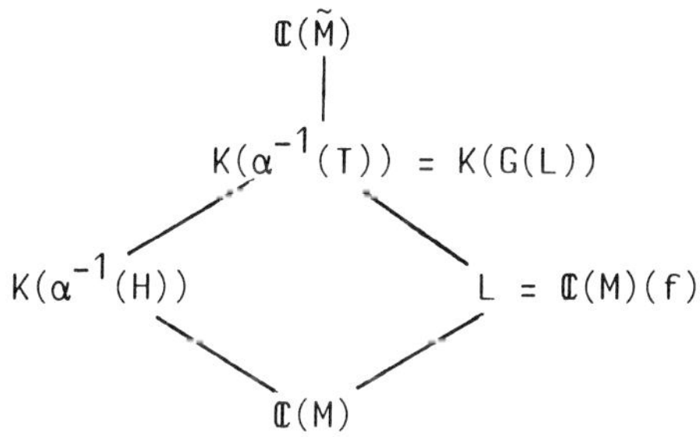

2.3 Abelian extensions

We follow the notations and the assumptions in section 2.1. (But the main results of this section turn out to be free from Condition 1.2.4 as will be shown.)

DEFINITION 2.3.1 A non-zero meromorphic function f on $\tilde{M} = \tilde{M}(D)$ is called a (resp. *strict*) *multiplicative function* if

$$\varphi^* f = \chi(\varphi) f \quad \text{for all} \quad \varphi \in G_{\tilde{\pi}},$$

where $\chi(\varphi) \in \mathbb{C}^* = \mathbb{C} - \{0\}$ (resp. $\chi(\varphi) \in \{z \in \mathbb{C} \mid |z| = 1\}$).

The multiple $\chi(\varphi)$ in the definition must satisfy the following condition as is easily seen:

$$\chi(\varphi_1\varphi_2) = \chi(\varphi_1)\chi(\varphi_2) \quad \text{for} \quad \varphi_1, \varphi_2 \in G_{\tilde{\pi}}.$$

Hence

$$\chi : \varphi \in G_{\tilde{\pi}} \longmapsto \chi(\varphi) \in \mathbb{C}^* = GL(1,\mathbb{C})$$

is a homomorphism. Thus a (resp. strict) multiplicative function is nothing but a non-degenerate representative matrix function ($r = 1$) with respect to the (resp. unitary) representation χ. χ is called the *multiplier of* f.

If f_1 and f_2 are (resp. strict) multiplicative functions with the multipliers χ_1 and χ_2, respectively, then f_1f_2 is a (resp. strict) multiplicative function with the multiplier

$$\chi_1\chi_2 : \varphi \in G_{\tilde{\pi}} \longmapsto \chi_1(\varphi)\chi_2(\varphi) \in \mathbb{C}^*.$$

Since $\mathbb{C}^*$ is abelian, a homomorphism $\chi : G_{\tilde{\pi}} \longrightarrow \mathbb{C}^*$ splits as follows:

$$\begin{array}{ccc} G_{\tilde{\pi}} & \longrightarrow & G_{\tilde{\pi}}/[G_{\tilde{\pi}},G_{\tilde{\pi}}] \\ & {\scriptstyle\chi}\searrow \quad \swarrow{\scriptstyle\hat{\chi}} & \\ & \mathbb{C}^* & \end{array}$$

We identify χ with $\hat{\chi}$.

LEMMA 2.3.2 Let $\gamma_1,\ldots,\gamma_s$ be the closed paths in M-B defined in section 1.2. Regarding them as elements of $H_1(M\text{-}B;\mathbb{Z})$,

$$G_{\tilde{\pi}}/[G_{\tilde{\pi}},G_{\tilde{\pi}}] \simeq H_1(M\text{-}B;\mathbb{Z})/(\mathbb{Z}(e_1\gamma_1) + \ldots + \mathbb{Z}(e_s\gamma_s)).$$

PROOF. Consider the projections

$$\alpha : \pi_1(M\text{-}B,\ p_o) \longrightarrow \pi_1(M\text{-}B,\ p_o)/[\pi_1,\pi_1] \simeq H_1(M\text{-}B;\mathbb{Z})$$

(Hurewicz isomorphism), and

$$\beta : \pi_1(M\text{-}B,\ p_o) \longrightarrow G_{\tilde{\pi}} = \pi_1(M\text{-}B,\ p_o)/J.$$

(J is the smallest normal subgroup of $\pi_1(M\text{-}B,\ p_o)$ which contains $\gamma_1^{e_1},\ldots,\gamma_s^{e_s}$.) Note first that

$$\beta([\pi_1,\pi_1]) = [G_{\tilde{\pi}},G_{\tilde{\pi}}].$$

Hence

$$\beta^{-1}([G_{\tilde{\pi}},G_{\tilde{\pi}}]) = [\pi_1,\pi_1].J.$$

Note second that

$$\alpha^{-1}(\mathbb{Z}(e_1\gamma_1)+\ldots+\mathbb{Z}(e_s\gamma_s)) = [\pi_1,\pi_1].J.$$

Thus

$$G_{\tilde{\pi}}/[G_{\tilde{\pi}},G_{\tilde{\pi}}] \simeq (\pi_1/J)/([\pi_1,\pi_1].J/J) \simeq \pi_1/([\pi_1,\pi_1].J)$$
$$= \pi_1/\alpha^{-1}(\mathbb{Z}(e_1\gamma_1)+\ldots+\mathbb{Z}(e_s\gamma_s)) \simeq H_1(M-B;\mathbb{Z})/(\mathbb{Z}(e_1\gamma_1)+\ldots+\mathbb{Z}(e_s\gamma_s)). \quad \square$$

We identify $G_{\tilde{\pi}}/[G_{\tilde{\pi}},G_{\tilde{\pi}}]$ with $H_1(M-B;\mathbb{Z})/(\mathbb{Z}(e_1\gamma_1)+\ldots+\mathbb{Z}(e_s\gamma_s))$ through the isomorphism in the lemma. We denote this additive group by A:

$$A = G_{\tilde{\pi}}/[G_{\tilde{\pi}},G_{\tilde{\pi}}] = H_1(M-B;\mathbb{Z})/(\mathbb{Z}(e_1\gamma_1)+\ldots+\mathbb{Z}(e_s\gamma_s)).$$

We note that $M-B = \bar{M}-\bar{B}$ and

$$H_1(\bar{M}-\bar{B};\ \mathbb{Z}) \simeq H^{2n-1}(\bar{M},\bar{B};\ \mathbb{Z}) \quad \text{(Lefschetz duality)}$$

(see, for example, Spanier [58]). Moreover, there is the following exact sequence:

$$\begin{array}{ccccccccc}
\longrightarrow H^{2n-2}(\bar{M};\mathbb{Z}) & \longrightarrow & H^{2n-2}(\bar{B};\mathbb{Z}) & \longrightarrow & H^{2n-1}(\bar{M},\bar{B};\mathbb{Z}) & \longrightarrow & H^{2n-1}(\bar{M};\mathbb{Z}) & \longrightarrow 0 \\
\wr\text{(Poincaré duality)} & & & & \wr \quad \circlearrowleft & & \wr\text{(Poincaré)} & \\
H_2(\bar{M};\mathbb{Z}) & & & & H_1(\bar{M}-\bar{B};\mathbb{Z}) & \xrightarrow{i_*} & H_1(\bar{M};\mathbb{Z}) &
\end{array}$$

where i_* is the homomorphism induced by the inclusion i: $M-B \hookrightarrow \bar{M}$. Hence

<u>LEMMA 2.3.3</u> There is the following exact sequence:

$$\longrightarrow H_2(\bar{M};\mathbb{Z}) \longrightarrow H^{2n-2}(\bar{B};\mathbb{Z}) \longrightarrow H_1(\bar{M}-\bar{B};\mathbb{Z}) \xrightarrow{i_*} H_1(\bar{M};\mathbb{Z}) \longrightarrow 0.$$

In particular, $H_1(\bar{M}-\bar{B};\mathbb{Z})$ is a finitely generated abelian group.

Following Pontryagin [50], a homomorphism

$$\chi : A \longrightarrow \{z \in \mathbb{C} \mid |z| = 1\}$$

is called a *character of* A. The set A* of all characters of A forms an abelian group in a natural way and is called the *dual group of* A. Under a suitable topology, A* is a topological abelian group. (The topology of A is the discrete topology.) The Pontryagin duality theorem says that

$$(A^*)^* \simeq A \quad \text{as topological groups.}$$

Let $M(M,D)$ (resp. $M_s(M,D)$) be the set of all (resp. strict) multiplicative functions. Under the product of functions, this is an abelian group (resp. a subgroup of $M(M,D)$). The correspondence

$$\begin{array}{ccc} M_s(M,D) & \longrightarrow & A^* \\ \cup & & \cup \\ f & \longmapsto & \chi = \text{the multiplier of } f \end{array}$$

is a homomorphism. By Theorem 2.2.1, this is surjective. The kernel is clearly the multiplicative group $\mathbb{C}(\dot{M})^* = \mathbb{C}(\dot{M}) - \{0\}$. Thus

<u>PROPOSITION 2.3.4</u> $M_s(M,D)/\mathbb{C}(M)^* \simeq A^*$ (as abelian groups).

Let f be a multiplicative function with the multiplier χ. Since

$$\varphi^*f = \chi(\varphi)f \quad \text{for all} \quad \varphi \in G_{\tilde{\pi}},$$

the divisor $(f) = D_0(f) - D_\infty(f)$ on $\tilde{M}$ of f is $G_{\tilde{\pi}}$-invariant. From this, it seems that (f) can be regarded as a divisor on M, but a strange phenomenon occurs at the points of $\tilde{\pi}^{-1}(B)$, the ramification locus of $\tilde{\pi}$. We explain this in the following.

For points $p \in B$ and $q \in \tilde{\pi}^{-1}(p)$, let W and U be as before open neighbourhoods of p and q with coordinate systems $(w_1,\ldots,w_n)$ and $(z_1,\ldots,z_n)$, respectively, such that $B \cap W = \{w_n = 0\}$ and

$$\tilde{\pi} : (z_1,\ldots,z_n) \in U \longmapsto (w_1,\ldots,w_n) = (z_1,\ldots,z_{n-1},z_n^e) \in W.$$

($p \in D_n$, say, and $e_n = e$.) By the proof of Theorem 2.2.1 and Lemma 2.2.8, f can be written as

$$f(z_1,\ldots,z_n) = z_n^a g(\tilde{\pi}(z_1,\ldots,z_n)) = w_n^{a/e} g(w_1,\ldots,w_n)$$

on U, where g is a meromorphic function on W and a is an integer such that $0 \leq a \leq e-1$. Hence the divisor (f) can be locally considered as

$$(f) = \frac{a}{e}(w_n) + (g) = \frac{a}{e} D_n + (g).$$

Thus the divisor (f) can be regarded as a $\mathbb{Q}$-divisor (i.e. a rational divisor) on M of the form

$$\frac{a_1}{e_1} D_1 + \ldots + \frac{a_s}{e_s} D_s + E,$$

where $0 \leq a_j \leq e_j-1$ $(1 \leq j \leq s)$ and E is a usual (integral) divisor on M. By the Remmert-Stein continuation theorem (see Narasimhan [45]), E can be continued to an (integral) divisor $\bar{E}$ on $\bar{M}$. Hence (f) can be regarded as a $\mathbb{Q}$-divisor on $\bar{M}$ of the form

$$\frac{a_1}{e_1}\bar{D}_1 + \dots + \frac{a_s}{e_s}\bar{D}_s + \bar{E}\ . \qquad \dots(*)$$

In general, we call a $\mathbb{Q}$-divisor on $\bar{M}$ of the form in (*) a *$\bar{D}$-divisor*. The set $\mathrm{Div}(\bar{M},\bar{D})$ of all $\bar{D}$-divisors forms an additive group.

A $\bar{D}$-divisor can be regarded as a singular (2n-2)-cycle with rational coefficients. We denote by $\mathrm{Div}_0^{\mathbb{Q}}(\bar{M},\bar{D})$ the subgroup of $\mathrm{Div}(\bar{M},\bar{D})$ of all $\bar{D}$-divisors which are $\mathbb{Q}$-homologous to zero.

The *Chern class of a $\bar{D}$-divisor* is defined as an element of $H^{1,1}(\bar{M},\mathbb{Q})$ as follows:

$$c(\frac{a_1}{e_1}\bar{D}_1 + \dots + \frac{a_s}{e_s}\bar{D}_s + \bar{E}) = \frac{a_1}{e_1}c([\bar{D}_1]) + \dots + \frac{a_s}{e_s}c([\bar{D}_s]) + c([\bar{E}]).$$

As is well known, the Chern class of a $\mathbb{Q}$-divisor $\hat{E}$ is Poincaré dual to $\hat{E}$. Hence

LEMMA 2.3.5 $\mathrm{Div}_0^{\mathbb{Q}}(\bar{M},\bar{D}) = \{\hat{E} \in \mathrm{Div}(\bar{M},\bar{D}) \mid c(\hat{E}) = 0 \in H^{1,1}(\bar{M},\mathbb{Q})\}$.

Since

$$(f_1f_2) = (f_1) + (f_2) \quad \text{for} \quad f_1,f_2 \in \mathcal{M}(M,D),$$

the correspondence

$$\rho : f \in \mathcal{M}(M,D) \longmapsto (f) \in \mathrm{Div}(\bar{M},\bar{D})$$

is a homomorphism.

THEOREM 2.3.6 (Weill [65]). $\rho(\mathcal{M}(M,D)) = \rho(\mathcal{M}_s(M,D)) = \mathrm{Div}_0^{\mathbb{Q}}(\bar{M},\bar{D})$.

For the proof of the theorem, we follow the discussion in de Rham-Kodaira [12] or Akizuki [1] (in Japanese).

Let

$$\hat{E} = \frac{b_1}{e_1}\bar{D}_1 + \dots + \frac{b_s}{e_s}\bar{D}_s + \bar{E}$$

be a $\bar{D}$-divisor, where b_j $(1 \le j \le s)$ are integers and $\bar{E}$ is an (integral) divisor on $\bar{M}$ whose support contains no $\bar{D}_j$. Let $\{W_\alpha\}$ be an open covering of $\bar{M}$ such that

$$\bar{D}_j \cap W = (g_{j\alpha})$$

and

$$\bar{E} \cap W_\alpha = (h_\alpha),$$

where $g_{j\alpha}$ and h_α are meromorphic functions on W_α. Consider the meromorphic 1-form

$$\eta_\alpha = \sum_{j=1}^{s} \frac{b_j}{e_j} \frac{dg_{j\alpha}}{g_{j\alpha}} + \frac{dh_\alpha}{h_\alpha} \quad \text{on } W_\alpha.$$

On $W_\alpha \cap W_\beta$, there are never vanishing holomorphic functions $g_{j\alpha\beta}$ and $h_{\alpha\beta}$ such that

$$g_{j\alpha} = g_{j\alpha\beta} g_{j\beta} \quad \text{and} \quad h_\alpha = h_{\alpha\beta} h_\beta.$$

Then

$$\eta_\alpha = \eta_\beta + \eta_{\alpha\beta} \quad \text{on} \quad W_\alpha \cap W_\beta,$$

where

$$\eta_{\alpha\beta} = \sum_{j=1}^{s} \frac{b_j}{e_j} \frac{dg_{j\alpha\beta}}{g_{j\alpha\beta}} + \frac{dh_{\alpha\beta}}{h_{\alpha\beta}} = \sum_{j=1}^{s} \quad d\log(g_{j\alpha\beta}) + d\log(h_{\alpha\beta})$$

is a closed holomorphic 1-form on $W_\alpha \cap W_\beta$.

Now, we recall the notion of currents. A *current* Θ of degree p on an open set W of $\bar{M}$ is, by definition, a continuous linear functional on the vector space $\Phi^{2n-p}(W)$ of all differentiable (2n-p)-forms on W with compact support, under a suitable topology. Θ can be locally written as

$$\Theta = \Sigma \vartheta_{j_1 \dots j_q; k_1 \dots k_r} dz^{j_1} \wedge \dots \wedge dz^{j_q} \wedge d\bar{z}^{k_1} \wedge \dots \wedge d\bar{z}^{k_r},$$

$(q+r = p)$, where $\vartheta_{j_1 \dots j_q; k_1 \dots k_r}$ are Schwartz' distributions. If every term is of type (q,r), then Θ is said to be *of type* (q,r). The operators $d, \partial, \bar{\partial}, d^*, \partial^*, \bar{\partial}^*$ (the adjoint operators), $\Delta, \square$, G (the Green operator), H (the harmonic projection), etc. can be extended to the operators on currents. Here, the metric on $\bar{M}$ is the Hodge metric induced by an inclusion $\bar{M} \subset \mathbb{P}^N$ and by the Fubini-Study metric on $\mathbb{P}^N$. The de Rham-Kodaira decomposition: $\Theta = H\Theta + \Delta G \Theta$ holds for a current Θ on $\bar{M}$. A fundamental theorem on currents asserts that a harmonic current Θ, $\Delta\Theta = 0$, is equal to a harmonic form (as currents). See de Rham [11] or Griffiths-Harris [22] for the rigorous definition and properties of currents.

Now, the meromorphic 1-form η_α on W_α defined above can be regarded as a (1,0)-current on W_α. We have on $W_\alpha \cap W_\beta$,

$$d\eta_\alpha = d\eta_\beta + d\eta_{\alpha\beta} = d\eta_\beta \quad \text{(as currents).}$$

Hence the global current

$$Z = d\eta_\alpha = d\eta_\beta$$

on $\bar{M}$ is defined. It can be shown that Z is a closed (1,1)-current.

On the other hand, the $\bar{D}$-divisor $\hat{E}$ can be regarded as a singular (2n-2)-cycle with rational coefficients, hence as a closed (1,1)-current on $\bar{M}$.

For the proof of the following lemma, see de Rham-Kodaira [12] or Griffiths-Harris [22].

<u>LEMMA 2.3.7</u> <u>(Poincaré-Lelong Equation)</u> $Z = 2\pi\sqrt{-1}\hat{E}$.

For the proof of the following lemma, see de Rham-Kodaira [12] or Akizuki [1].

<u>LEMMA 2.3.8</u> If a (p,0)-current Θ defined on an open set W of a compact Kähler manifold is closed, then Θ is harmonic, and hence is equal to a closed holomorphic (p,0)-form (as currents) on W.

Now, suppose that a $\bar{D}$-divisor $\hat{E}$ is written as

$$\hat{E} = (f)$$

for a multiplicative function f. Note that the meromorphic 1-form

$$\eta = df/f$$

on $\tilde{M}$ is $G_{\tilde{\pi}}$-invariant, so can be regarded as a meromorphic 1-form on M, hence on $\bar{M}$. It is clear that $\eta - \eta_\alpha$ is a closed holomorphic 1-form on W . Hence

$$d\eta = Z \quad \text{(as currents).}$$

By Lemma 2.3.7, $\hat{E}$ is exact as a current. Hence $\hat{E}$ is $\mathbb{Q}$-homologous to zero as a singular (2n-2)-cycle with rational coefficients. Thus

$$\rho(M(M,d)) \subset \mathrm{Div}_0^{\mathbb{Q}}(\bar{M},\bar{D}).$$

Conversely, suppose that a $\bar{D}$-divisor $\hat{E}$ is $\mathbb{Q}$-homologous to zero. Then $Z = 2\pi\sqrt{-1}\hat{E}$ is exact. Let ω be any holomorphic 1-form on $\bar{M}$. Put

$$\eta = 2\bar{\partial}^*GZ + \omega = 2\pi\sqrt{-1}(2\bar{\partial}^*G\hat{E}) + \omega.$$

η is then a (1,0)-current on $\bar{M}$. Note that $Z = \bar{\partial}\eta_\alpha$ on W_α, (since Z is of type (1,1)) and that $HZ = 0$, (since Z is exact). Note also that

$$\partial G = G\partial,\ \bar{\partial} G = G\bar{\partial},\ \partial\bar{\partial} = -\bar{\partial}\partial,\ \partial\bar{\partial}^* = -\bar{\partial}^*\partial.$$

Hence we have

$$\begin{aligned} d\eta &= d(2\bar{\partial}^* GZ) = (\partial+\bar{\partial})(2\bar{\partial}^* GZ) \\ &= 2\partial\bar{\partial}^* G\bar{\partial}\eta_\alpha + 2\bar{\partial}\bar{\partial}^* G\bar{\partial}\eta_\alpha \\ &= 2\bar{\partial}^*\bar{\partial} G\partial\eta_\alpha + 2\bar{\partial}\bar{\partial}^* G\bar{\partial}\eta_\alpha \\ &= 2\Box G d\eta_\alpha = \Delta GZ = Z - HZ = Z. \end{aligned}$$

If there is another (1,0)-current η' on $\bar{M}$ such that $d\eta' = Z$, then $d(\eta'-\eta) = 0$. By Lemma 2.3.8,

$$\eta' - \eta = \omega'$$

is a holomorphic 1-form on $\bar{M}$. Thus

$$\eta' = \eta + \omega' = 2\bar{\partial}^* GZ + \omega + \omega'.$$

Since $d(\eta-\eta_\alpha) = d\eta - d\eta_\alpha = Z - Z = 0$, $\eta = \eta_\alpha$ is a closed holomorphic 1-form on W_α by Lemma 2.3.8. We may assume that W_α is a polydisc with respect to a coordinate system. Then there is a never vanishing holomorphic function u_α on W_α such that

$$\eta - \eta_\alpha = d\log(u_\alpha).$$

The support of the current $Z = 2\pi\sqrt{-1}\hat{E}$ is the support $\mathrm{Supp}(\hat{E})$ of the $\bar{D}$-divisor $\hat{E}$. Hence

$$d\eta = Z = 0 \quad \text{on} \quad M - \mathrm{Supp}(\hat{E}).$$

By Lemma 2.3.8 again, η is a closed holomorphic 1-form on $\bar{M} - \mathrm{Supp}(\hat{E})$. Put

$$f(q) = c \exp \int_\delta \eta \quad \text{for } q \in \tilde{M} - \tilde{\pi}^{-1}(\mathrm{Supp}(\hat{E})),$$

where c is a non-zero constant and q corresponds to the path δ which starts from p_o and does not pass through $\mathrm{Supp}(\hat{E})$.

Since $\eta - \eta_\alpha = d\log(u_\alpha)$ on W_α, we have

$$f = c\, u_\alpha h_\alpha \prod_{j=1}^{s} g_{j\alpha}^{b_j/e_j} \quad \text{on } W_\alpha - \mathrm{Supp}(\hat{E}).$$

Hence f can be extended to a meromorphic function f on $\tilde{M}$. We have clearly

$$(f) = \hat{E}$$

and

$$(\gamma^*f)(q) = f(\gamma(q)) = c \exp \int_{\gamma+\delta} \eta = (\exp \int_\gamma \eta) f(q),$$

where γ is an integral 1-cycle on $\bar{M}$ such that

$$\gamma \cap \mathrm{Supp}(\hat{E}) = \emptyset.$$

Hence f is a multiplicative function with the multiplier

$$\chi : \gamma \in H_1(M-B;\mathbb{Z}) \longmapsto \exp \int_\gamma \eta \in \mathbb{C}^*.$$

Next, since $\hat{E}$ is $\mathbb{Q}$-homologous to zero, there is a singular (2n-1)-chain C on $\bar{M}$ with rational coefficients such that

$$\partial C = \hat{E}, \quad \text{that is,} \quad dC = \hat{E} \quad \text{(as currents).}$$

Then

$$\int_\gamma \eta = 2\pi\sqrt{-1} \int_\gamma 2\bar{\partial}^*GdC + \int_\gamma \omega.$$

For the proof of the following lemma, see de Rham-Kodaira [12] or Akizuki [1].

<u>LEMMA 2.3.9</u> $I(\gamma,C) = \int_\gamma 2\bar{\partial}^*GdC - \int_C H\gamma,$

where $I(\gamma,C)$ is the intersection number of γ and C, and $H\gamma$ is the harmonic part of γ (as a current).

By the lemma, we have

$$\chi(\gamma) = \exp\{2\pi\sqrt{-1}(I(\gamma,C) + \int_C H\gamma) + \int_\gamma \omega\} .$$

If $\omega = 0$, then $|\chi(\gamma)| = 1$. In fact

$$\overline{(\int_C H\gamma)} = \int_C \overline{H\gamma} = \int_C H\bar{\gamma} = \int_C H\gamma$$

is real. Conversely, if $|\chi(\gamma)| = 1$, then

$$\mathrm{Re}(\int_\gamma \omega) = 0.$$

But

<u>LEMMA 2.3.10</u> If $\mathrm{Re}(\int_\gamma \omega) = 0$ for any integral 1-cycle γ on $\bar{M}-\bar{B}$, hence on $\bar{M}$ (by Lemma 2.3.3), then $\omega = 0$.

PROOF. Consider the function

$$h(p) = \int_{p_0}^{p} (\omega+\bar{\omega})/2.$$

By the assumption, $h(p)$ does not depend on the path from p_0 to p. Hence h is a differentiable function on $\bar{M}$. We have $\partial\bar{\partial}h = 0$. Hence $\mathrm{Re}(h)$ and $\mathrm{Im}(h)$ are harmonic functions on $\bar{M}$, so are constants. Thus

$$(\omega+\bar{\omega})/2 = dh = 0.$$

This means that $\omega = 0$. □

Hence we have the following theorem, which completes the proof of Theorem 2.3.6.

THEOREM 2.3.11 (Kodaira) If a $\bar{D}$-divisor $\hat{E}$ is $\mathbb{Q}$-homologous to zero, then

$$f(q) = c \exp 2\pi\sqrt{-1} \int_{\delta} 2\bar{\partial}{*}G\hat{E}$$

($q = [\delta]$, $c \in \mathbb{C}^*$) is the unique (up to non-zero constants) strict multiplicative function such that $(f) = \hat{E}$. In this case, the multiplier $\chi = \chi_{\hat{E}}$ of f is uniquely determined by $\hat{E}$ and is given by

$$\chi_{\hat{E}}(\gamma) = \exp 2\pi\sqrt{-1}(I(\gamma,C) + \int_C H\gamma)$$

for $\gamma \in H_1(\bar{M}-\bar{B};\mathbb{Z})$, where C is a singular $(2n-1)$-chain on $\bar{M}$ with rational coefficients such that $\partial C = \hat{E}$.

Two $\bar{D}$-divisors $\hat{E}_1$ and $\hat{E}_2$ are said to be *linearly equivalent*, $\hat{E}_1 \sim \hat{E}_2$, if $\hat{E}_1 - \hat{E}_2$ is an (integral) divisor and is linearly equivalent to zero in the usual sense. This is an equivalence relation. We denote by $\hat{E}_1/\sim$ the equivalence class to which $\hat{E}_1$ belongs.

In particular, $\hat{E}_1 \sim 0$ means that $\hat{E}_1$ is itself an (integral) divisor and is linearly equivalent to zero.

As a corollary to Theorem 2.3.11, we have the following generalization of the famous theorem of Abel:

THEOREM 2.3.12 (Kodaira). Let $\hat{E}$ be a $\bar{D}$-divisor on $\bar{M}$. Then $\hat{E} \sim 0$ if and only if (1) $\hat{E}$ is $\mathbb{Q}$-homologous to zero and (2) $I(\gamma,C) + \int_C H\gamma \equiv 0 \pmod 1$

for any integral 1-cycle γ on $\bar{M}-\bar{B}$, where C is a singular $(2n-1)$-chain on $\bar{M}$ with rational coefficients such that $\partial C = \hat{E}$.

We denote by $\mathrm{Div}_0^{\mathbb{Q}}(\bar{M},\bar{D})/\sim$ the set of all equivalence classes of $\bar{D}$-divisors which are $\mathbb{Q}$-homologous to zero. This is an additive group, called the *$\bar{D}$-divisor class group*.

PROPOSITION 2.3.13 The surjective homomorphism

$$\rho : f \in M_s(M,d) \longmapsto (f) \in \mathrm{Div}_0^{\mathbb{Q}}(\bar{M},\bar{D})$$

induces the isomorphism

$$M_s(M,D)/\mathbb{C}(M)^* \simeq \mathrm{Div}_0^{\mathbb{Q}}(\bar{M},\bar{D})/\sim.$$

PROOF. Put

$$\mathrm{Div}_\ell(\bar{M}) = \{\hat{E} \mid \hat{E} \sim 0\}.$$

By Theorem 2.3.11, we have the following exact sequences:

$$1 \longrightarrow \mathbb{C}^* \longrightarrow M_s(M,D) \longrightarrow \mathrm{Div}_0^{\mathbb{Q}}(\bar{M},\bar{D}) \longrightarrow 0$$

and

$$1 \longrightarrow \mathbb{C}^* \longrightarrow \mathbb{C}(M)^* \longrightarrow \mathrm{Div}_\ell(\bar{M}) \longrightarrow 0.$$

Hence

$$M_s(M,D)/\mathbb{C}(M)^* \simeq \mathrm{Div}_0^{\mathbb{Q}}(\bar{M},\bar{D})/\mathrm{Div}_\ell(\bar{M}). \quad \square$$

This, together with Proposition 2.3.3, implies

COROLLARY 2.3.14 $\mathrm{Div}_0^{\mathbb{Q}}(\bar{M},\bar{D})/\sim \; \simeq A^*$ (as abelian groups).

We introduce a topology in $\mathrm{Div}_0^{\mathbb{Q}}(\bar{M},\bar{D})/\sim$ so that this is isomorphic to A^* as topological abelian groups. Then we have

PROPOSITION 2.3.15 The abelian groups $\mathrm{Div}_0^{\mathbb{Q}}(\bar{M},\bar{D})/\sim$ and $A = H_1(\bar{M}-\bar{B};\mathbb{Z})/(\mathbb{Z}(e_1\gamma_1)+\ldots+\mathbb{Z}(e_s\gamma_s))$ are dual as topological abelian groups with respect to the pairing induced by

$$(\hat{E},\gamma) \in \mathrm{Div}_0^{\mathbb{Q}}(\bar{M},\bar{D}) \times H_1(M-B;\mathbb{Z})$$
$$\longmapsto \chi_{\hat{E}}(\gamma) = \exp 2\pi\sqrt{-1}\,\{I(\gamma,C) + \int_C H\gamma\} \in \{z \in \mathbb{C} \mid |z| = 1\},$$

where C is a singular $(2n-1)$-chain on $\bar{M}$ with rational coefficients such that $\partial C = \hat{E}$.

Now, by Schur's lemma, an irreducible unitary representation of a finite abelian group G is a character of G. Hence, by Proposition 2.2.12,

<u>PROPOSITION 2.3.16</u> Any finite abelian extension K of $\mathbb{C}(M)$ with $K \subset \mathbb{C}(\tilde{M})$ can be obtained by attaching some strict multiplicative functions to $\mathbb{C}(M)$.

Our goal in this section is the following theorem, which corresponds to a fundamental result in class field theory (see Weil [66]).

<u>THEOREM 2.3.17</u> There is a one-to-one correspondence between the set of all finite abelian extensions K of $\mathbb{C}(M)$ with $K \subset \mathbb{C}(\tilde{M})$ and the set of all finite subgroups S of $\mathrm{Div}_0^{\mathbb{Q}}(\bar{M},\bar{D})/\sim$. The correspondence $K \longrightarrow S = S(K)$ is given by

$$S(K) = \{(f)/\sim \mid f \text{ is a strict multiplicative function in } K\}.$$

Moreover, (1) $S(K) \simeq \mathrm{Gal}(K/\mathbb{C}(M))$ and (2) $K_1 \subset K_2$ if and only if $S(K_1) \subset S(K_2)$.

<u>PROOF</u>. We use the results in Pontryagin [50]. The set of all subgroups Γ of A and the set of all subgroups S of $\mathrm{Div}_0^{\mathbb{Q}}(M,D)/\sim$ are of one-to-one correspondence. The correspondence is given by

$$\Gamma \longrightarrow S(\Gamma) = \{\hat{E}/\sim \mid \chi_{\hat{E}}(\gamma) = 1 \quad \text{for all} \quad \gamma \in A\},$$

and

$$S \longrightarrow \Gamma(S) = \{\gamma \in A \mid \chi_{\hat{E}}(\gamma) = 1 \text{ for all } \quad \hat{E}/\sim \in S\}.$$

Moreover, we have

$$S(\Gamma) \simeq (A/\Gamma)^*$$

and

$$\Gamma(S) \simeq ((\mathrm{Div}_0^{\mathbb{Q}}(\bar{M},\bar{D})/\sim)/S)^*.$$

In particular, for a subgroup H of $G_{\tilde{\pi}}$ with $[G_{\tilde{\pi}},G_{\tilde{\pi}}] \subset H$, the quotient group

$$\Gamma = H/[G_{\tilde{\pi}},G_{\tilde{\pi}}]$$

is of finite index in $A = G_{\tilde{\pi}}/[G_{\tilde{\pi}},G_{\tilde{\pi}}]$ if and only if $S(\Gamma)$ is a finite subgroup of $\mathrm{Div}_0^{\mathbb{Q}}(\bar{M},\bar{D})/\sim$. In this case,

$$S(\Gamma) \simeq S(\Gamma)^* \simeq A/\Gamma \simeq G_{\tilde{\pi}}/H \simeq \mathrm{Gal}(K(H)/\mathbb{C}(M)),$$

where K(H) is the field corresponding to H (see section 2.1).

Now our correspondence is given by

$$K \longmapsto H = G(K) \longmapsto \Gamma = H/[G_{\tilde{\pi}},G_{\tilde{\pi}}] \longmapsto S(\Gamma) = S(K). \quad \square$$

Note that $S(K) \simeq \mathrm{Gal}(K/\mathbb{C}(M))$ is *not* canonical, while the dual group $S(K)^*$ is canonically isomorphic to $\mathrm{Gal}(K/\mathbb{C}(M))$.

The above theorem can be rewritten as follows:

THEOREM 2.3.18 Let $\bar{D}_j$ $(1 \le j \le s)$ be distinct irreducible hypersurfaces of a projective manifold $\bar{M}$. Let e_j $(1 \le j \le s)$ be positive integers and put $\bar{D} = e_1\bar{D}_1 + \ldots + e_s\bar{D}_s$. Then there is a one-to-one correspondence between the set of all isomorphism classes of finite abelian coverings $\bar{\pi} : \bar{X} \longrightarrow \bar{M}$ which branches at most at $\bar{D}$ and the set of all finite subgroups S of $\mathrm{Div}_0^{\mathbb{Q}}(\bar{M},\bar{D})/\sim$. The correspondence $\bar{\pi} \longmapsto S = S(\bar{\pi})$ satisfies that (1) $S(\bar{\pi}) \simeq G_{\bar{\pi}}$ and (2) $\bar{\pi}_1 \le \bar{\pi}_2$ if and only if $S(\bar{\pi}_1) \subset S(\bar{\pi}_2)$.

PROOF. By Theorems 2.1.3 and 1.2.7, there is a one-to-one correspondence between the set of all finite abelian extensions K of $\mathbb{C}(M)$ with $K \subset \mathbb{C}(\tilde{M})$ and the set of all isomorphism classes of finite abelian coverings $\pi : X \longrightarrow M$ which branches at most at D. Let $\bar{\pi} : \bar{X} \longrightarrow \bar{M}$ be the $\mathbb{C}(X)$-normalization of $\bar{M}$. Then the correspondence $\pi \longmapsto \bar{\pi}$ is one-to-one. $\square$

Note again that $S(\bar{\pi}) \simeq G_{\bar{\pi}}$ is *not* canonical, while the dual group $S(\bar{\pi})^*$ is canonically isomorphic to $G_{\bar{\pi}}$.

REMARK 2.3.19 *Theorem 2.3.18 is independent of Condition 1.2.4.* In fact, let e_j' be the smallest positive integer such that $\gamma_j^{e_j'} \in J$. Then, by lemma 1.2.3 e_j' divides e_j and Condition 1.2.4 is satisfied for (M,D'), where

$$D' = e_1'D_1 + \ldots + e_s'D_s.$$

Hence we have a maximal covering

$$\tilde{\pi} : \tilde{M} = \tilde{M}(D') \longrightarrow M.$$

Note that any covering $\pi : X \longrightarrow M$ which branches at most at D is necessarily a covering which branches at most at D'. In fact, by Lemma 1.2.2, $J \subset H = \pi_*(\pi_1(X-\pi^{-1}(B), q_0))$. If π is locally written as

$$\pi : (z_1,\ldots,z_n) \longmapsto (w_1,\ldots,w_n) = (z_1,\ldots,z_{n-1},z_n^d),$$

where $\{w_n = 0\} = D_n$ locally, say, then $\gamma_n^d \in H$ and $\gamma_n^{d'} \notin H$ for $0 < d' < d$. If $e_n' = ad+b$ where $0 \le b < d$, then $\gamma_n^b = \gamma_n^{e_n'}\gamma_n^{-ad} \in H$. Hence $b = 0$, so d divides e_n'.

Hence, in order to show our assertion, it suffices to show that

$$\mathrm{Div}_o^{\mathbb{Q}}(\bar{M},\bar{D}) = \mathrm{Div}_o^{\mathbb{Q}}(\bar{M},\bar{D}'),$$

where

$$\bar{D}' = e_1'\bar{D}_1 + \ldots + e_s'\bar{D}_s.$$

Let

$$\hat{E} = \frac{b_1}{e_1}\bar{D}_1 + \ldots + \frac{b_s}{e_s}\bar{D}_s + \bar{E}$$

be an element of $\mathrm{Div}_o^{\mathbb{Q}}(\bar{M},\bar{D})$. Using the notations in the proof of Theorem 2.3.6, for an integer ν,

$$\exp \int_{\nu\gamma_k} \eta = \exp \int_{\nu\gamma_k} \Big(\sum_{j=1}^{s} \frac{b_j}{e_j} \frac{dg_{j\alpha}}{g_{j\alpha}} + \frac{dh_\alpha}{h_\alpha} + \frac{du_\alpha}{u_\alpha}\Big)$$

$$= \zeta_k^{\nu b_k} \qquad \ldots(*)$$

where $\zeta_k = \exp 2\pi\sqrt{-1}/e_k$. Note that, as an element of $H_1(\bar{M}-\bar{B};\mathbb{Z})$, $e_k'\gamma_k$ can be written as

$$e_k'\gamma_k = c_1e_1\gamma_1 + \ldots + c_se_s\gamma_s$$

for some integers c_j $(1 \le j \le s)$. Hence

$$\exp \int_{e_k'\gamma_k} \eta = \exp \int_{c_1e_1\gamma_1+\ldots+c_se_s\gamma_s} \Big(\sum_{j=1}^{s} \frac{b_j}{e_j} \frac{dg_{j\alpha}}{g_{j\alpha}} + \frac{dh_\alpha}{h_\alpha} + \frac{du_\alpha}{u_\alpha}\Big) = 1.$$

Hence, by (*), we have

$$e_k'b_k \equiv 0 \pmod{e_k}.$$

Thus b_k can be divided by e_k/e_k' for $1 \le k \le s$. Hence, writing $b_k = b_k'(e_k/e_k')$,

$$\hat{E} = \frac{b_1'}{e_1'}\bar{D}_1 + \ldots + \frac{b_s'}{e_s'}\bar{D}_s + \bar{E} \in \mathrm{Div}_o^{\mathbb{Q}}(\bar{M},\bar{D}').$$

Thus

$$\mathrm{Div}_o^{\mathbb{Q}}(\bar{M},\bar{D}) = \mathrm{Div}_o^{\mathbb{Q}}(\bar{M},\bar{D}').$$

(Also, $f = c \exp \int_{\delta} \eta$ (= $cu_{\alpha} h_{\alpha} \cdot \prod_{j=1}^{s} g_{j\alpha}^{b'_j/e'_j}$ locally) is a meromorphic function on $\tilde{M}(D')$.)

From this remark, we can find, conversely, a criterion for the existence of a finite abelian covering $\bar{\pi} : \bar{X} \longrightarrow \bar{M}$ which branches *at* $\bar{D}$ as follows.

By the proof of Theorem 1.2.7, there is a finite abelian covering $\pi : X \longrightarrow M$ which branches at D if and only if there is a normal subgroup H of $G_{\tilde{\pi}}$ (= the covering transformation group of $\tilde{\pi} : M = M(D') \longrightarrow M$ such that (1) H is of finite index in $G_{\tilde{\pi}}$ and $G_{\tilde{\pi}}/H$ is abèlian and (2) $\gamma_j^d \in H$ implies $d \equiv 0 \pmod{e_j}$ for $1 \leq j \leq s$. (This also implies that $D' = D$.)

Considering the correspondence

$$H \longrightarrow \Gamma = H/[G_{\tilde{\pi}}, G_{\tilde{\pi}}] \longrightarrow S = S(\Gamma)$$

in the proof of Theorem 2.3.17, we assert that $d\gamma_j \in \Gamma$ if and only if

$$1 = \chi_{\hat{E}}(d\gamma_j) = \exp \int_{d\gamma_j} \eta = \zeta_j^{db_j}$$

for all $\hat{E}/\sim = (\frac{b_1}{e_1}\bar{D}_1 + \ldots + \frac{b_s}{e_s}\bar{D}_s + \bar{E})/\sim \in S$, where $\zeta_j = \exp 2\pi\sqrt{-1}/e_j$.

Thus, as a solution to a Fenchel's type problem:

<u>THEOREM 2.3.20</u> Let $\bar{D}_j$ $(1 \leq j \leq s)$ be distinct irreducible hypersurfaces of a projective manifold $\bar{M}$. Let e_j $(1 \leq j \leq s)$ be positive integers and put $\bar{D} = e_1\bar{D}_1 + \ldots + e_s\bar{D}_s$. Then there is a finite abelian covering $\bar{\pi} : \bar{X} \longrightarrow \bar{M}$ which branches *at* $\bar{D}$ if and only if there is a finite subgroup S of $\mathrm{Div}_0^{\mathbb{Q}}(\bar{M}, \bar{D})/\sim$ such that, for every j, there is

$$\hat{E}/\sim = \hat{E}(j)/\sim = (\frac{b_1}{e_1}\bar{D}_1 + \ldots + \frac{b_s}{e_s}\bar{D}_s + \bar{E})/\sim$$

in S such that $(b_j, e_j) = 1$ (coprime).

It should be noted again that *this theorem is independent of Condition 1.2.4.* (Rather, if the condition of this theorem is satisfied, then Condition 1.2.4 on (M,D) is necessarily satisfied.)

In Theorem 2.3.18, if $\bar{B} = \bar{D}_1 \cup \ldots \cup \bar{D}_s$ is empty, then $\bar{M} = M$, $\tilde{\pi} : \tilde{M} \longrightarrow M$ is the universal covering, $A = H_1(M; \mathbb{Z})$ and

$$A^* \simeq \mathrm{Div}_o^{\mathbb{Q}}(M)/\sim \; = \{E \mid E \text{ is an (integral) divisor on } M$$
$$\text{such that } c_{\mathbb{Z}}([E]) \text{ is a torsion in } H^2(M;\mathbb{Z})\}/\sim$$
$$\simeq \mathrm{Pic}_t(M),$$

where $c_{\mathbb{Z}} : \mathrm{Pic}(M) \longrightarrow H^{1,1}(M, \mathbb{Z})$ is the (integral) Chern class and

$$\mathrm{Pic}_t(M) = \{L \in \mathrm{Pic}(M) \mid c_{\mathbb{Z}}(L) \text{ is a torsion in } H^2(M;\mathbb{Z})\}.$$

Thus, as a corollary to Theorem 2.3.18,

THEOREM 2.3.21 Let M be a projective manifold. Then there is a one-to-one correspondence between the set of all isomorphism classes of unramified finite abelian coverings $\pi : X \longrightarrow M$ and the set of all finite subgroups S of $\mathrm{Pic}_t(M)$. The correspondence $\pi \longmapsto S = S(\pi)$ satisfies that (1) $S(\pi) \simeq G_\pi$ and (2) $\pi_1 \leq \pi_2$ if and only if $S(\pi_1) \subset S(\pi_2)$.

In general, the structure of the group $\mathrm{Div}_o^{\mathbb{Q}}(M,D)/\sim$ can be seen as follows: We put

$$\Lambda(\bar{M},\bar{D}) = \{(b_1(\mathrm{mod}\ e_1),\ldots,b_s(\mathrm{mod}\ e_s)) \in (\mathbb{Z}/e_1\mathbb{Z}) \times\ldots\times (\mathbb{Z}/e_s\mathbb{Z}) \;\Big|\; \frac{b_1}{e_1} j_*c([\bar{D}_1])+\ldots+ \frac{b_s}{e_s} j_*c([\bar{D}_s]) \in j_*(H^{1,1}(\bar{M},\mathbb{Z}))\},$$

where $j_* : H^{1,1}(\bar{M},\mathbb{Z}) \longrightarrow H^{1,1}(\bar{M},\mathbb{Q})$ is the homomorphism induced by $j : \mathbb{Z} \hookrightarrow \mathbb{Q}$. $\Lambda(\bar{M},\bar{D})$ is then a finite abelian group. (We put $\Lambda(\bar{M},\bar{D}) = 0$ if $\bar{B}$ is empty.) We put also

$$\mathrm{Pic}_o(\bar{M}) = \{L \in \mathrm{Pic}(\bar{M}) \mid c_{\mathbb{Z}}(L) = 0 \in H^{1,1}(\bar{M},\mathbb{Z})\}$$

and

$$T^2(\bar{M}) = \text{the torsion subgroup of } H^2(\bar{M};\mathbb{Z}).$$

($\mathrm{Pic}_o(\bar{M})$ is an abelian variety of dimension $q = b_1/2$ (the irregularity of $\bar{M}$).)

PROPOSITION 2.3.22 There are the following exact sequences:

$$0 \longrightarrow \mathrm{Pic}_t(\bar{M}) \longrightarrow \mathrm{Div}_o^{\mathbb{Q}}(\bar{M},\bar{D})/\sim \longrightarrow \Lambda(\bar{M},\bar{D}) \longrightarrow 0$$
$$0 \longrightarrow \mathrm{Pic}_o(\bar{M}) \longrightarrow \mathrm{Pic}_t(\bar{M}) \longrightarrow T^2(\bar{M}) \longrightarrow 0.$$

PROOF. To any D-divisor class

$$\hat{E}/\sim \;=\; \left(\frac{b_1}{e_1}\bar{D}_1 + \ldots + \frac{b_s}{e_s}\bar{D}_s + \bar{E}\right)/\sim,$$

we associate the point

$$(b_1 (\mathrm{mod}\ e_1), \ldots, b_s (\mathrm{mod}\ e_s)) \in (\mathbb{Z}/e_1\mathbb{Z}) \times \ldots \times (\mathbb{Z}/e_s\mathbb{Z}).$$

This is well defined and

$$\frac{b_1}{e_1} c([\bar{D}_1]) + \ldots + \frac{b_s}{e_s} c([\bar{D}_s]) = c([-\bar{E}]) \in j_*(H^{1,1}(\bar{M},\mathbb{Z})).$$

Conversely, if the last equation holds, then

$$\frac{b_1}{e_1}\bar{D}_1 + \ldots + \frac{b_s}{e_s}\bar{D}_s + \bar{E} \in \mathrm{Div}_o^{\mathbb{Q}}(\bar{M},\bar{D}).$$

Now, $b_j \equiv 0 \pmod{e_j}$ for $1 \leq j \leq s$ if and only if $\hat{E}$ is an (integral) divisor. The group of all (integral) divisor classes $\hat{E}/\sim$ such that $c([\hat{E}]) = 0$ is isomorphic to $\mathrm{Pic}_t(\bar{M})$.

The exactness of the second sequence is obvious. □

Next, we examine the above discussion by examples.

In the rest of this section, we write M, B, D_j, D, *etc. instead of* $\bar{M}$, $\bar{B}$ $\bar{D}_j$, $\bar{D}$, *etc. for simplicity.*

EXAMPLE 2.3.23 Let D_j $(1 \leq j \leq s)$ be (distinct) irreducible hypersurfaces of a projective manifold M. Put $B = D_1 \cup \ldots \cup D_s$ and $D = D_1 + \ldots + D_s$, (that is, D is a reduced positive divisor). Suppose that there is a positive integer e and a (not necessarily positive) divisor E such that $[eE] = [D]$. Consider the $\mathbb{Q}$-divisor

$$\hat{E} = \frac{1}{e} D - E.$$

Then $\hat{E}$ is clearly an element of $\mathrm{Div}_o^{\mathbb{Q}}(M,eD)$. Note that

$$a\left(\frac{1}{e} D - E\right) \sim b\left(\frac{1}{e} D - E\right) \quad \text{(linearly equivalent)}$$

if and only if

$$a \equiv b \pmod{e}.$$

Hence

$$S = \mathbb{Z}\hat{E}/\sim\ = \{\frac{a}{e} D - aE \mid 0 \le a \le e-1\}$$

is a finite cyclic subgroup of $\mathrm{Div}_o^{\mathbb{Q}}(M,eD)/\sim$. The cyclic covering

$$\pi : X \longrightarrow M$$

corresponding to S is nothing but the covering given in 1° of section 1.4.

<u>EXAMPLE 2.3.24</u> Let $D_1,\ldots,D_s$ $(s \ge 2)$ be linearly equivalent reduced positive divisors of a projective manifold M, any two of which do not have common component. Put $D = D_1 + \ldots + D_s$. For an integer e (≥ 2),

$$\hat{E}_{12} = \frac{1}{e} D_1 - \frac{1}{e} D_2$$

is an element of $\mathrm{Div}_o^{\mathbb{Q}}(M,eD)$. Moreover, $a\hat{E}_{12}$ and $b\hat{E}_{12}$ are linearly equivalent if and only if $a \equiv b \pmod e$. Hence

$$S_{12} = \mathbb{Z}\hat{E}_{12}/\sim\ = \{a\hat{E}_{12} \mid 0 \le a \le e-1\}$$

is a finite cyclic subgroup of $\mathrm{Div}_o^{\mathbb{Q}}(M,eD)/\sim$.

In a similar way, we can define finite cyclic subgroups

$$S_{23},\ldots,S_{n-1,n},S_{n1}$$

of $\mathrm{Div}_o^{\mathbb{Q}}(M,eD)/\sim$. Put

$$S = S_{12} + S_{23} + \ldots + S_{n-1,n} + S_{n1}.$$

Note that

$$\hat{E} = m_1(\frac{1}{e} D_1 - \frac{1}{e} D_2) + \ldots + m_s(\frac{1}{e} D_s - \frac{1}{e} D_1) \in S$$

is linearly equivalent to zero if and only if

$$m_1 \equiv \ldots \equiv m_s \pmod e.$$

Hence

$$S \simeq (\mathbb{Z}/(e\mathbb{Z}))^s/\Delta,$$

where

$$\Delta = \{(m,\ldots,m) \mid m \in \mathbb{Z}/e\mathbb{Z}\}.$$

The order of S is e^{s-1}. The finite abelian covering

$$\pi : X \longrightarrow M$$

corresponding to S is nothing but the covering in 6° of section 1.4.

EXAMPLE 2.3.25 Put $M = \mathbb{P}^n$ and D_j = an irreducible hypersurface of $\mathbb{P}^n$ of degree d_j ($1 \leq j \leq s$). Put $D = e_1D_1+\ldots+e_sD_s$. Note that

$$c_{\mathbb{Z}} : \mathrm{Pic}(\mathbb{P}^n) \simeq H^2(\mathbb{P}^n;\mathbb{Z}) = \mathbb{Z} \quad \text{(identified)}.$$

Hence

$$\mathrm{Pic}_o(\mathbb{P}^n) = \mathrm{Pic}_t(\mathbb{P}^n) = 0.$$

Hence

$$\mathrm{Div}_o^{\mathbb{Q}}(\mathbb{P}^n,D)/\sim \; \simeq \Lambda(\mathbb{P}^n,D) =$$
$$\{(b_1,\ldots,b_s) \in (\mathbb{Z}/E_1\mathbb{Z})\times\ldots\times(\mathbb{Z}/e_s\mathbb{Z}) \;\Big|\; \frac{b_1d_1}{e_1}+\ldots+\frac{b_sd_s}{e_s} \in \mathbb{Z}\}.$$

This additive group is a finite group, to which there corresponds an abelian covering

$$\pi : X \longrightarrow \mathbb{P}^n$$

which branches at most at D. Note that the additive group

$$A = \{c_1\gamma_1+\ldots+c_s\gamma_s \mid c_j \in \mathbb{Z},\; d_1\gamma_1+\ldots+d_s\gamma_s = e_1\gamma_1 = \ldots = e_s\gamma_s = 0\}$$

and $\Lambda(\mathbb{P}^n,D)$ are dual to each other, by the pairing

$$(c_1\gamma_1+\ldots+c_s\gamma_s), \quad (b_1,\ldots,b_s)) \in A \times \Lambda(\mathbb{P}^n,D)$$
$$\longmapsto \chi = \exp 2\pi\sqrt{-1}\left(\frac{b_1c_1}{e_1} + \ldots + \frac{b_sc_s}{c_s}\right).$$

By theorem 2.3.20, $\pi : X \longrightarrow \mathbb{P}^n$ branches at D if and only if, for every j, there is a point $(b_1,\ldots,b_s) \in \Lambda(\mathbb{P}^n,D)$ such that $(b_j,e_j) = 1$.

This condition is equivalent to the condition in Theorem 1.3.41. In this case, $\pi : X \longrightarrow \mathbb{P}^n$ is a maximal abelian covering which branches at D, (see Example 1.3.39).

EXAMPLE 2.3.26 Let $D_1,\ldots,D_s$ ($s \geq 2$) be linearly equivalent irreducible hypersurfaces of a projective manifold M. Put

$$D = e_1D_1 + \ldots + e_sD_s.$$

Consider the $\mathbb{Q}$-divisor

$$\hat{E} = \frac{b_1}{e_1}D_1 + \ldots + \frac{b_s}{e_s}D_s - \left(\frac{b_1}{e_1} + \ldots + \frac{b_s}{e_s}\right)D_1,$$

where $b_1,\ldots,b_s$ are integers such that

$$\frac{b_1}{e_1} + \dots + \frac{b_s}{e_s} \in \mathbb{Z}.$$

Then $\hat{E}$ is clearly an element of $\mathrm{Div}_o^{\mathbb{Q}}(M,D)$. Note that $\hat{E}$ is linearly equivalent to zero if and only if

$$b_1 \equiv 0 \pmod{e_1}, \dots, b_s \equiv 0 \pmod{e_s}.$$

Hence

$$S = \left\{\frac{b_1}{e_1} D_1 + \dots + \frac{b_s}{e_s} D_s - \left(\frac{b_1}{e_1} + \dots + \frac{b_s}{e_s}\right) D_1 \,\middle|\, b_j \in \mathbb{Z}, \frac{b_1}{e_1} + \dots + \frac{b_s}{e_s} \in \mathbb{Z}\right\} \Big/ \sim$$

$$\simeq \left\{(b_1, \dots, b_s) \in (\mathbb{Z}/e_1\mathbb{Z}) \times \dots \times (\mathbb{Z}/e_s\mathbb{Z}) \,\middle|\, \frac{b_1}{e_1} + \dots + \frac{b_s}{e_s} \in \mathbb{Z}\right\}$$

is a finite subgroup of $\mathrm{Div}_o^{\mathbb{Q}}(M,D)/\sim$. The finite abelian covering

$$\pi : X \longrightarrow M$$

corresponding to S branches at most at D.

The dual group S* to S is, as Example 2.3.25,

$$S^* = \{c_1\gamma_1 + \dots + c_s\gamma_s \mid c_j \in \mathbb{Z}, \ \gamma_1 + \dots + \gamma_s = e_1\gamma_1 = \dots = e_s\gamma_s = 0\}.$$

Hence, as Lemma 1.3.33, $\pi : X \longrightarrow M$ branches at D if and only if, for every j ($1 \leq j \leq s$), e_j divides

$$\langle e_1, \dots, \check{e}_j, \dots, e_s \rangle.$$

Thus

THEOREM 2.3.27 Let $D_1, \dots, D_s$ ($s \geq 2$) be linearly equivalent irreducible hypersurfaces of a projective manifold M. Let e_j ($1 \leq j \leq s$) be positive integers such that, for every j, e_j divides $\langle e_1, \dots, \check{e}_j, \dots, e_s \rangle$. Then there exists a finite abelian covering $\pi : X \longrightarrow M$ which branches at $D = e_1 D_1 + \dots + e_s D_s$.

The covering $\pi : X \longrightarrow M$ in this theorem can also be constructed using a discussion similar to that in the proof of Theorem 1.5.2, with respect to the linear system $|D_1|$.

EXAMPLE 2.3.28 Let M be a compact Riemann surface of genus g, $B = \{p_1, \dots, p_s\}$ and $D = e_1(p_1) + \dots + e_s(p_s)$. ($(p_j)$ is the point divisor of the point $p_j \in M$.) Then there is the following exact sequence

$$0 \longrightarrow \mathrm{Jac}(M) \longrightarrow \mathrm{Div}_o^{\mathbb{Q}}(M,D)/\sim \longrightarrow \Lambda(e_1, \dots, e_s) \longrightarrow 0,$$

where Jac(M) is the Jacobian variety of M, and

$$\Lambda(e_1,\ldots,e_s) = \left\{(b_1,\ldots,b_s) \in (\mathbb{Z}/e_1\mathbb{Z}) \times \ldots \times (\mathbb{Z}/e_s\mathbb{Z}) \;\middle|\; \frac{b_1}{e_1} + \ldots + \frac{b_s}{e_s} \in \mathbb{Z}\right\}$$

(= 0 if B is empty).

Moreover, this exact sequence splits, that is,

$$\mathrm{Div}_o^{\mathbb{Q}}(M,D)/\!\sim\; \simeq \mathrm{Jac}(M) \oplus \Lambda(e_1,\ldots,e_s) \qquad \text{(a direct sum)}.$$

In fact, under the notation of Example 1.3.42,

$$H_1(M-B;\mathbb{Z}) \,/\, (\mathbb{Z}(e_1\gamma_1) + \ldots + \mathbb{Z}(e_s\gamma_s))$$

is isomorphic to

$$\mathbb{Z}\alpha_1 \oplus \mathbb{Z}\beta_1 \oplus \ldots \oplus \mathbb{Z}\alpha_g \oplus \mathbb{Z}\beta_g \oplus A \qquad \text{(a direct sum)},$$

where

$$A = \{c_1\gamma_1+\ldots+c_s\gamma_s \mid c_j \in \mathbb{Z},\quad \gamma_1+\ldots+\gamma_s = e_1\gamma_1 = \ldots = e_s\gamma_s = 0\}.$$

Hence its dual group $\mathrm{Div}_o^{\mathbb{Q}}(M,D)/\!\sim$ is isomorphic to

$$\mathrm{Jac}(M) \oplus \Lambda(e_1,\ldots,e_s).$$

We identify $\mathrm{Div}_o^{\mathbb{Q}}(M,D)/\!\sim$ with $\mathrm{Jac}(M) \oplus \Lambda(e_1,\ldots,e_s)$ through the isomorphism. Let

$$\rho : \mathrm{Div}_o^{\mathbb{Q}}(M,D)/\!\sim\; = \mathrm{Jac}(M) \oplus \Lambda(e_1,\ldots,e_s) \longrightarrow \mathrm{Jac}(M)$$

be the projection. Let S be a finite subgroup of $\mathrm{Div}_o^{\mathbb{Q}}(M,D)/\!\sim$ which satisfies the condition of Theorem 2.3.20. Then

$$\rho(S) \oplus \Lambda(e_1,\ldots,e_s)$$

is also a finite subgroup which satisfies the condition of Theorem 2.3.20 and contains S. Since $\Lambda(e_1,\ldots,e_s)$ is dual to A,

THEOREM 2.3.29 Let M be a compact Riemann surface and $D = e_1(p_1) + \ldots + e_s(p_s)$ be a positive divisor on M. Then there is a finite abelian covering $\pi : X \longrightarrow M$ which branches at D if and only if, for every j $(1 \leq j \leq s)$, e_j divides $\langle e_1,\ldots,\check{e}_j,\ldots,e_s\rangle$.

Finally, we move D and consider various $\mathrm{Div}_o^{\mathbb{Q}}(M,D)/\!\sim$'s on a fixed projective manifold M. We first note that

$$\mathrm{Div}_o^{\mathbb{Q}}(M,D) \subset \mathrm{Div}_o^{\mathbb{Q}}(M,D'),$$

for $D = e_1D_1+\ldots+e_sD_s$ and $D' = e_1'D_1+\ldots+e_s'D_s$ such that e_j divides e_j' for $1 \leq j \leq s$.

Consider the additive group

$$\mathrm{Div}_o^{\mathbb{Q}}(M) = \{r_1D_1+\ldots+r_sD_s \mid r_j \in \mathbb{Q},\ D_j \text{ is a prime divisor } (1 \leq j \leq s),$$
$$r_1j_*c([D_1])+\ldots+r_sj_*c([D_s]) = 0 \in H^{1,1}(M,\mathbb{Q})\}.$$

This is a subgroup of the group of all rational divisors on M. We have clearly

$$\mathrm{Div}_o^{\mathbb{Q}}(M) = \bigcup_D \mathrm{Div}_o^{\mathbb{Q}}(M,D)$$

where $\cup$ runs over all positive divisors D on M.

Two elements $\hat{E}_1$ and $\hat{E}_2$ of $\mathrm{Div}_o^{\mathbb{Q}}(M)$ are said to be *linearly equivalent*, $\hat{E}_1 \sim \hat{E}_2$, if $\hat{E}_1 - \hat{E}_2$ is an integral and principal divisor. By Theorem 2.3.18, we conclude

THEOREM 2.3.30 Let M be a projective manifold. Then there is a one-to-one correspondence $\pi \longmapsto S = S(\pi)$ between the set of all isomorphism classes of finite abelian branched coverings $\pi : X \longrightarrow M$ and the set of all finite subgroups S of $\mathrm{Div}_o^{\mathbb{Q}}(M)/\sim$. The correspondence satisfies (1) $S(\pi) \simeq G_\pi$ and (2) $\pi_1 \leq \pi_2$ if and only if $S(\pi_1) \subset S(\pi_2)$.

Note that the isomorphism $S(\pi) \simeq G_\pi$ in the theorem is not canonical, while there is a canonical isomorphism

$$S(\pi)^* \simeq G_\pi,$$

where $S(\pi)^*$ is the dual group of $S(\pi)$.

Using the one-to-one correspondence

$$\pi \longmapsto K = \mathbb{C}(X),$$

this theorem can be written as

THEOREM 2.3.30' Let M be a projective manifold. Then there is a one-to-one correspondence $K \longrightarrow S = S(K)$ between the set of all (isomorphism classes of) finite abelian extensions $K/\mathbb{C}(M)$ and the set of all finite subgroups S of $\mathrm{Div}_o^{\mathbb{Q}}(M)/\sim$. The correspondence satisfies (1) $S(K) \simeq \mathrm{Gal}(K/\mathbb{C}(M))$ and (2) $K_1 \subset K_2$ if and only if $S(K_1) \subset S(K_2)$.

REMARK 2.3.31 The class field theory for fields of algebraic functions of one variable asserts the dual version of this theorem, using the *generalized Jacobian variety*, (see Serre [55]).

3 Weil-Tōyama theory

3.1 Matrix D-divisors and D-vector bundles

The purpose of this chapter is to extend the results for abelian extensions in section 2.3 to those for Galois extensions. We follow Weil [64], Tōyama [62] and Grothendieck [23].

We follow the notations and the assumptions in section 2.1, which we repeat here: Let $\bar{M}$ be an n-dimensional projective manifold and $\bar{B}$ be a hypersurface of $\bar{M}$. ($\bar{B}$ may possibly be empty.) Let

$$\bar{B} = \bar{D}_1 \cup \ldots \cup \bar{D}_s$$

be the decomposition of $\bar{B}$ into irreducible components. For positive integers $e_1,\ldots,e_s$, we consider the positive divisor

$$\bar{D} = e_1\bar{D}_1+\ldots+e_s\bar{D}_s$$

on $\bar{M}$. Put

$$M = \bar{M} - \mathrm{Sing}\bar{B},$$

$$B = \bar{B} - \mathrm{Sing}\bar{B}, \quad D_j = \bar{D}_j - \mathrm{Sing}\bar{B},$$

and

$$D = e_1D_1+\ldots+e_sD_s.$$

Suppose that Condition 1.2.4 is satisfied for the pair (M,D). Then we have a maximal covering

$$\tilde{\pi} : \tilde{M} = \tilde{M}(D) \longrightarrow M$$

which branches at D. M is then the quotient space

$$M = \tilde{M}/G_{\tilde{\pi}},$$

where $G_{\tilde{\pi}}$ is the covering transformation group of $\tilde{\pi}$, acting on $\tilde{M}$ properly discontinuously. We write $\tilde{G}$ instead of $G_{\tilde{\pi}}$ for convenience:

$$\tilde{G} = G_{\tilde{\pi}}.$$

DEFINITION 3.1.1 A fibre space $S \longrightarrow \tilde{M}$ is called a $\tilde{G}$-*fibre space* if $\tilde{G}$ acts also on S and the action is compatible with the projection $S \longrightarrow \tilde{M}$. In particular, if S is a sheaf, then such S is called a $\tilde{G}$-*sheaf*.

For a $\tilde{G}$-sheaf S, put

$H^0(\tilde{M},\tilde{G};S)$ = the set of all $\tilde{G}$-invariant sections of S,

$\tilde{\pi}_*^{\tilde{G}}(S)$ = the associated sheaf on M to the presheaf

$$W \longmapsto H^0(\tilde{\pi}^{-1}(W),\tilde{G};S).$$

Then it is clear that $H^0(M,\tilde{\pi}_*^{\tilde{G}}(S))$ is naturally isomorphic to $H^0(\tilde{M},\tilde{G};S)$.

Note that $\tilde{\pi}_*^{\tilde{G}}(\mathcal{O}_{\tilde{M}}) = \mathcal{O}_M$ and that, if a $\tilde{G}$-sheaf S is a coherent $\mathcal{O}_{\tilde{M}}$-module, then $\tilde{\pi}_*^{\tilde{G}}(S)$ is a coherent $\mathcal{O}_M$-module.

Next, let S be a $\tilde{G}$-sheaf on $\tilde{M}$ of (not necessarily commutative) groups, such that the group action is compatible with $\tilde{G}$-action, i.e.

$$\varphi(ab) = \varphi(a)\varphi(b) \quad \text{and} \quad \varphi(1) = 1 \quad \text{for } \varphi \in \tilde{G}.$$

Let $\mathcal{U} = \{U_\alpha\}$ be an open covering of $\tilde{M}$. Let $\xi_{\alpha\beta}$ be a $\tilde{G}$-invariant section of S on $U_\alpha \cap U_\beta$ such that

$$\xi_{\alpha\alpha} = 1 \quad \text{on} \quad U_\alpha,$$

$$\xi_{\alpha\beta}\xi_{\beta\gamma}\xi_{\gamma\alpha} = 1 \quad \text{on} \quad U_\alpha \cap U_\beta \cap U_\gamma,$$

and for $\varphi \in \tilde{G}$, if $\varphi(U_\alpha) = U_\gamma$ and $\varphi(U_\beta) = U_\delta$, then there are sections $\omega_\alpha(\varphi)$ on U_γ and $\omega_\beta(\varphi)$ on U_δ such that

$$\varphi(\xi_{\alpha\beta}) = \omega_\alpha(\varphi)\xi_{\gamma\delta}\omega_\beta(\varphi)^{-1}.$$

The collection $\{\xi_{\alpha\beta}\}$ of such $\xi_{\alpha\beta}$'s is called a *cocycle*. We denote by $Z^1(\mathcal{U},\tilde{G};S)$ the set of all cocycles. This is not necessarily a group. Two elements $\{\xi_{\alpha\beta}\}$ and $\{\xi'_{\alpha\beta}\}$ are said to be *equivalent*, $\{\xi_{\alpha\beta}\} \sim \{\xi'_{\alpha\beta}\}$, if there is a collection $\{\eta_\alpha\}$ of sections η_α on U_α such that

$$\xi'_{\alpha\beta} = \eta_\alpha\xi_{\alpha\beta}\eta_\beta^{-1} \quad \text{on} \quad U_\alpha \cap U_\beta.$$

This is an equivalence relation. Put

$$H^1(\mathcal{U},\tilde{G};S) = Z^1(\mathcal{U},\tilde{G};S)/\sim,$$

the set of all equivalence classes. Also put

$$H^1(\tilde{M},\tilde{G};S) = \lim_{\mathcal{U}} H^1(\mathcal{U},\tilde{G};S).$$

<u>LEMMA 3.1.2</u> Let S be a $\tilde{G}$-sheaf of groups and J be a $\tilde{G}$-subsheaf of S. Then (1) the sheaf S/J of left cosets is a $\tilde{G}$-sheaf and (2) there is the following exact sequence:

$$1 \longrightarrow H^0(\tilde{M},\tilde{G};\mathcal{J}) \longrightarrow H^0(\tilde{M},\tilde{G};\mathcal{J})$$
$$\longrightarrow H^0(\tilde{M},\tilde{G};\mathcal{S}/\mathcal{J}) \xrightarrow{\delta} H^1(\tilde{M},\tilde{G};\mathcal{J}),$$

where δ is defined as follows: $\delta(\zeta) = \xi_\alpha \xi_\beta^{-1}$, where $\zeta \in H^0(\tilde{M},\tilde{G};\mathcal{S}/\mathcal{J})$ is expressed on U_α by a section ξ_α of $\mathcal{S}$.

Note that $H^1(\tilde{M},\tilde{G};\mathcal{J})$ is not necessarily a group. The last exactness in the lemma says that, if elements ζ and ζ' of $H^0(\tilde{M},\tilde{G};\mathcal{S}/\mathcal{J})$ are expressed on U_α by ξ_α and ξ'_α, respectively, then $\delta(\zeta) = \delta(\zeta')$ if and only if there are $\xi \in H^0(\tilde{M},\tilde{G};\mathcal{S})$ and a section η_α on U_α of $\mathcal{J}$ such that

$$\xi'_\alpha = \eta_\alpha \xi_\alpha \xi.$$

This is clear. The other assertions of the lemma are trivial.

Now, let $\mathcal{M}^*_{\tilde{M}}(r)$ (resp. $\mathcal{O}^*_{\tilde{M}}(r)$) be the sheaf on $\tilde{M}$ of $(r \times r)$-matrix valued meromorphic (resp. holomorphic) functions F such that det(F) is not identically zero (resp. is never vanishing). Then $\mathcal{M}^*_{\tilde{M}}(r)$ is a $\tilde{G}$-sheaf of groups and $\mathcal{O}^*_{\tilde{M}}(r)$ is its $\tilde{G}$-subsheaf. Put

$$\mathcal{D}_{\tilde{M}}(r) = \mathcal{M}^*_{\tilde{M}}(r)/\mathcal{O}^*_{\tilde{M}}(r)$$

and call it the *sheaf of (germs of) (r x r)-matrix* D-*divisors*. An element of $H^0(\tilde{M},\tilde{G};\ \mathcal{D}_{\tilde{M}}(r))$ is called an *(r x r)-matrix* D-*divisor* (or simply a *matrix* D-*divisor*). Grothendieck [23] called an element of $H^1(\tilde{M},\tilde{G};\mathcal{O}^*_{\tilde{M}}(r))$ an (isomorphism class of) $\tilde{G}$-vector bundles on $\tilde{M}$ of rank r, but we call it an (*isomorphism class of*) D-*vector bundles on* M *of rank* r in this book.

In a similar way to $\mathcal{M}^*_{\tilde{M}}(r)$ (resp. $\mathcal{O}^*_{\tilde{M}}(r)$), we can define the sheaf $\mathcal{M}^*_M(r)$ (resp. $\mathcal{O}^*_M(r)$) on M. Then we have

$$\tilde{\pi}_*^{\tilde{G}}(\mathcal{M}^*_{\tilde{M}}(r)) = \mathcal{M}^*_M(r) \quad \text{and} \quad \tilde{\pi}_*^{\tilde{G}}(\mathcal{O}^*_{\tilde{M}}(r)) = \mathcal{O}^*_M(r).$$

Hence

$$H^0(\tilde{M},\tilde{G};\ \mathcal{O}^*_{\tilde{M}}(r)) \simeq H^0(M,\mathcal{O}^*_M(r)) \simeq GL(r,\mathbb{C})$$

and

$$H^0(\tilde{M},\tilde{G};\mathcal{M}^*_{\tilde{M}}(r)) \simeq H^0(M,\mathcal{M}^*_M(r)).$$

We put

$$\mathrm{Div}(M,D,r) = H^0(\tilde{M},\tilde{G};\mathcal{D}_{\tilde{M}}(r))$$

and

$$V(M,D,r) = H^1(\tilde{M},\tilde{G},\mathcal{O}^*_{\tilde{M}}(r)).$$

By Lemma 3.1.2,

LEMMA 3.1.3 There is the following exact sequence:

$$1 \longrightarrow GL(r,\mathbb{C}) \longrightarrow H^0(M,\mathcal{M}^*_M(r)) \longrightarrow Div(M,D,r) \xrightarrow{\delta} V(M,D,r).$$

For a matrix D-divisor $\hat{E}$, we denote $\delta(\hat{E})$ by $[\hat{E}]$ and call in the *D-vector bundle associated to* $\hat{E}$.

PROBLEM Is the map δ in Lemma 3.1.3 surjective?

Note that a matrix D-divisor $\hat{E}$ is expressed on U_α by a matrix valued meromorphic function F_α such that $\det(F_\alpha)$ is not identically zero. We may write

$$\hat{E} = \{F_\alpha\}.$$

The exactness of the sequence in Lemma 3.1.3 says that $[\hat{E}] = [\hat{E}']$ if and only if there is a matrix valued meromorphic function P on M with non-zero $\det(P)$ and a matrix valued holomorphic function Q_α on U_α with never vanishing $\det(Q_\alpha)$ such that

$$F'_\alpha = Q_\alpha F_\alpha P \quad \text{for all } \alpha,$$

where $\hat{E} = \{F_\alpha\}$ and $\hat{E}' = \{F'_\alpha\}$. Since $\hat{E} = \{Q_\alpha F_\alpha\}$, we may write this last relation simply

$$E' = EP.$$

DEFINITION 3.1.4 $(r \times r)$-matrix D-divisors $\hat{E}$ and $\hat{E}'$ are said to be *linearly equivalent*, $\hat{E} \sim \hat{E}'$, if $[\hat{E}] = [\hat{E}']$, that is, if there is an $(r \times r)$ matrix valued meromorphic function P on M with non-zero $\det(P)$ such that $\hat{E}' = \hat{E}P$. An equivalence class is called a *matrix D-divisor class*.

This is an equivalence relation, which generalizes the usual one.

Now, we study matrix D-divisors more closely. Let $\hat{E} = \{F_\alpha\}$ be a matrix D-divisor.

If U_α does not intersect with the ramification locus $\tilde{\pi}^{-1}(B)$ of $\tilde{\pi}$, then we may assume that (1) U_α is a connected component of $\tilde{\pi}^{-1}(W_\alpha)$, where $W_\alpha = \tilde{\pi}(U_\alpha)$ and (2) $\tilde{\pi}$ maps every connected component $\tilde{\pi}^{-1}(W_\alpha)$ biholomorphically onto W_α. If $U_\gamma = \varphi(U_\alpha)$ for $\varphi \in \tilde{G}$, then U_γ is a connected component of $\tilde{\pi}^{-1}(W_\alpha)$ and there is a matrix valued holomorphic function Q_α on U_α with never vanishing $\det(Q_\alpha)$ such that

$$\varphi^* F_\gamma \quad (= F_\gamma . \varphi) = Q_\alpha F_\alpha .$$

By substituting $(\varphi^{-1})^*(Q_\alpha^{-1})F_\gamma$ for F_γ, we may assume that

$$\varphi^* F_\gamma = F_\alpha .$$

We may then regard F_α as a matrix valued meromorphic function on W_α through $\tilde{\pi} : U_\alpha \longrightarrow W_\alpha$. $M - B$ is covered by such W_α's:

$$M - B = \bigcup_{\alpha \in A} W_\alpha .$$

Hence, *on* $M - B$, $\hat{E}$ can be regarded as a collection $\{F_\alpha\}$, where F_α is a matrix valued meromorphic function on W_α, with non-zero $\det(F_\alpha)$, such that

$$F_{\alpha\beta} = F_\alpha F_\beta^{-1} \qquad \ldots(1)$$

is a matrix valued holomorphic function on $W_\alpha \cap W_\beta$ with never vanishing $\det(F_{\alpha\beta})$.

If U_ν intersects with $\tilde{\pi}^{-1}(B)$, then we may assume that (1) U_ν is a connected component of $\tilde{\pi}^{-1}(W_\nu)$, where $W_\nu = \tilde{\pi}(U_\nu)$, (2) W_ν satisfies condition (4) of Definition 1.1.1 and (3) U_ν and W_ν have coordinate systems $(z_1,\ldots,z_n)$ and $(w_1,\ldots,w_n)$, respectively, such that

$$\tilde{\pi} : (z_1,\ldots,z_n) \in U_\nu \longmapsto (w_1,\ldots,w_n) = (z_1,\ldots,z_{n-1},z_n^e) \in W_\nu .$$

($\{w_n = 0\} = D_n \cap W_\nu$, say, and $e = e_n$.)

Consider the biholomorphic map

$$\sigma_\nu : (z_1,\ldots,z_n) \in U_\nu \longmapsto (z_1,\ldots,z_{n-1},\zeta z_n) \in U_\nu$$

($\zeta = \exp(2\pi\sqrt{-1}/e)$) of U_ν onto itself. Then $\tilde{\pi}\sigma_\nu = \tilde{\pi}$. Since $\tilde{\pi}$ is a Galois covering, σ_ν can be uniquely extended to a covering transformation $\sigma_\nu \in \tilde{G}$. There is then a matrix valued holomorphic function S_ν on U_ν with never vanishing $\det(S_\nu)$ such that

$$\sigma_\nu^* F_\nu \quad (= F_\nu . \sigma_\nu) = S_\nu F_\nu ,$$

that is

$$F_\nu(z_1,\ldots,z_{n-1},\zeta z_n) = S_\nu(z_1,\ldots,z_n)F_\nu(z_1,\ldots,z_n).$$

We regard F_ν and S_ν as matrix valued functions of $(w_1,\ldots,w_{n-1},\sqrt[e]{w_n})$ on W_ν:

$$F_\nu = F_\nu(z_1,\ldots,z_n) = F_\nu(w_1,\ldots,w_{n-1},\sqrt[e]{w_n})$$

and

$$S_\nu = S_\nu(z_1,\dots,z_n) = S_\nu(w_1,\dots,w_{n-1}, \sqrt[e]{w_n})$$

such that

$$F_\nu(w_1,\dots,w_{n-1}, \zeta\sqrt[e]{w_n}) = S_\nu(w_1,\dots,w_{n-1},\sqrt[e]{w_n})F_\nu(w_1,\dots,w_{n-1},\sqrt[e]{w_n}) \qquad \dots(2)$$

$(\zeta = \exp 2\pi\sqrt{-1}/e)$.

We cover B by such W_ν's:

$$B \subset \bigcup_{\nu\in N} W_\nu.$$

Then M is covered by $\{W_\alpha\}_{\alpha\in A} \cup \{W_\nu\}_{\nu\in N}$, where $\{W_\alpha\}_{\alpha\in A}$ is the open covering of M - B defined above. We may write

$$\hat{E} = \{F_\alpha\} \cup \{F_\nu\}.$$

For $\alpha \in A$ and $\nu \in N$ with $W_\alpha \cap W_\nu \neq \emptyset$, (taking $\{W_\alpha\}_{\alpha\in A}$ sufficiently refined), we may assume that

$$F_{\alpha\nu} = F_\alpha F_\nu^{-1} \quad \text{and} \quad F_{\nu\alpha} = F_\nu F_\alpha^{-1} = F_{\alpha\nu}^{-1} \qquad \dots(3)$$

are matrix valued holomorphic functions on $W_\alpha \cap W_\nu$ with never vanishing $\det(F_{\alpha\nu})$ and $\det(F_{\nu\alpha})$.

For $\nu,\mu \in N$ with $W_\nu \cap W_\mu \neq \emptyset$,

$$F_{\nu\mu} = F_{\nu\mu}(w_1,\dots,w_{n-1},\sqrt[e]{w_n}) = F_\nu F_\mu^{-1} \qquad \dots(4)$$

is a matrix valued holomorphic function of $(w_1,\dots,w_{n-1},\sqrt[e]{w_n})$ on $W_\nu \cap W_\mu$ with never vanishing $\det(F_{\nu\mu})$, which satisfies

$$F_{\nu\mu}(w_1,\dots,w_{n-1}, \zeta\sqrt[e]{w_n})$$
$$= S_\nu(w_1,\dots,w_{n-1}, \sqrt[e]{w_n})F_{\nu\mu}(w_1,\dots,w_{n-1}, \sqrt[e]{w_n})S_\nu(w_1,\dots,w_{n-1}, \sqrt[e]{w_n})^{-1} \qquad \dots(5)$$

where $(w_1,\dots,w_n)$ is a local coordinate system on W_μ. (It can be easily shown that $\sigma_\nu = \sigma_\mu$ on $U_\nu \cap U_\mu$.)

Conversely, if matrix valued meromorphic functions F_α on W_α for all $\alpha \in A$ and

$$F_\nu = F_\nu(w_1,\dots,w_{n-1}, \sqrt[e]{w_n})$$

of $(w_1,\dots,w_{n-1}, \sqrt[e]{w_n})$ on W_ν for all $\nu \in N$ are given, and if they satisfy the conditions (1) - (4) above, then they define a matrix D-divisor

$$\hat{E} = \{F_\alpha\} \cup \{F_\nu\}.$$

The D-vector bundle $[\hat{E}]$ associated to $\hat{E}$ is then given by transition matrices appeared in (1) - (4) above:

$$[\hat{E}] = \{F_{\alpha\beta}\} \cup \{F_{\alpha\nu}\} \cup \{F_{\nu\alpha}\} \cup \{F_{\nu\mu}\} \ . \qquad \ldots(6)$$

In general, a D-vector bundle V on M can be regarded as a collection of transition matrices as in (6) which satisfy the compatibility conditions: $F_{\alpha\beta}F_{\beta\gamma} = F_{\alpha\gamma}$, etc and (5) above. Note that V is a usual holomorphic vector bundle on M - B, but may not be on whole M. (If $\bar{B}$ is empty, then V is a holomorphic vector bundle on $\bar{M} = M$. Moreover, V can always be written as $V = [\hat{E}]$ for a matrix divisor $\hat{E}$ on $\bar{M} = M$, by Nakano's theorem (Theorem 2.2.5).)

We can thus define (r x r)-matrix D-divisors and D-vector bundles of rank r as above, *without referring to* $\tilde{M}$ *or* $\tilde{G}$.

For r = 1, (1 x 1)-matrix D-divisors are nothing but $\bar{D}$-divisors defined in section 2.3.

For a matrix D-divisor $\hat{E} = \{F_\alpha\} \cup \{F_\nu\}$, put

$$\det(\hat{E}) = \{\det(F_\alpha)\} \cup \{\det(F_\nu)\}$$

and

$$\det[\hat{E}] = [\det(\hat{E})].$$

Then $\det(\hat{E})$ is a D-divisor (= a $\bar{D}$-divisor) on M. $\det[\hat{E}]$ is the D-line bundle associated with $\det(\hat{E})$ whose transition functions are $(\det(F_\alpha))(\det(F_\beta))^{-1}$, etc.

3.2 Representations and flat D-vector bundles

We follow the notation and the assumptions in the previous section. Let

$$R : \tilde{G} \longrightarrow GL(r,\mathbb{C})$$

be a representation of $\tilde{G}$. By Theorem 2.2.1, there is a non-degenerate representative matrix function F with respect to R, that is, an (r x r)-matrix valued meromorphic function F on $\tilde{M} = \tilde{M}(D)$ with non-zero det(F) such that

$$\varphi^*F = R(\varphi)F \quad \text{for all } \varphi \in \tilde{G}.$$

From this relation, it is clear that F defines a matrix D-divisor

$$(F) = \{F_\alpha\} \cup \{F_\nu\},$$

where $F_\alpha = F|U_\alpha$ and $F_\nu = F|U_\nu$, ($\alpha \in A$ and $\nu \in N$), under the notations in section 3.1. ($S_\nu = R(\sigma_\nu)$ is a constant function in this case.) (F) is called the *matrix D-divisor of the representative matrix function* F.

The D-vector bundle [(F)] associated to (F) is called the *flat D-vector bundle associated to* F. On M - B, (F) is equal to the flat vector bundle S_R which corresponds to the representation R, (see section 2.2). Recall that the bundle space of S_R is the quotient space

$$((\tilde{M} - \tilde{\pi}^{-1}(B)) \times \mathbb{C}^r)/\tilde{G}.$$

The action of $\tilde{G}$ is naturally extended to that on $\tilde{M} \times \mathbb{C}^r$:

$$(\varphi,p,\xi) \in \tilde{G} \times \tilde{M} \times \mathbb{C}^r \longmapsto (\varphi(p),R(\varphi)\xi) \in \tilde{M} \times \mathbb{C}^r.$$

The quotient space

$$V_R = (\tilde{M} \times \mathbb{C}^r)/\tilde{G}$$

may have singularity and is a normal complex space in general. There is the surjective holomorphic map

$$\tau : V_R \longrightarrow M,$$

which is induced from the projection

$$(p,\xi) \in \tilde{M} \times \mathbb{C}^r \longmapsto p \in \tilde{M}.$$

We may identify [(F)] with the fibre space $\tau : V_R \longrightarrow M$. We sometimes write V_R instead of [(F)] and call it the *flat D-vector bundle associated to* R. Note that

$$S_R = V_R \mid (\tilde{M} - \tilde{\pi}^{-1}(D))$$

as fibre spaces.

If F_1 is another non-degenerate representative matrix function with respect to R, then, by Lemma 2.2.8, there is a matrix valued meromorphic function P on M with non-zero det(P) such that

$$F_1 = FP.$$

This implies that

$$(F_1) = (F)P.$$

Hence

$$[(F_1)] = [(F)],$$

(see Definition 3.1.4).

Next, let

$$R' : \tilde{G} \longrightarrow GL(r,\mathbb{C})$$

be another representation of $\tilde{G}$ and F' be a non-degenerate representative matrix function with respect to R'.

Suppose that $[(F')] = [(F)]$, that is, there is a matrix valued meromorphic function P on M with non-zero det(P) such that

$$(F') = (F)P = (FP).$$

Then there are matrix valued holomorphic functions Q_α and Q_ν on U_α and U_ν, respectively, $(\alpha \in A,\ \nu \in N)$, with never vanishing $\det(Q_\alpha)$ and $\det(Q_\nu)$, such that

$$F' = Q_\alpha FP \quad \text{on} \quad U_\alpha,$$

and

$$F' = Q_\nu FP \quad \text{on} \quad U_\nu.$$

Then

$$Q = Q_\alpha = Q_\nu = F'P^{-1}F^{-1}$$

is a matrix valued holomorphic function on $\tilde{M}$ with never vanishing det(Q). Thus

$$F' = QFP.$$

Then, for any $\varphi \in \tilde{G}$,

$$\begin{aligned} R'(\varphi) &= (\varphi^*F')F'^{-1} \\ &= (\varphi^*Q)(\varphi^*F)(\varphi^*P)P^{-1}F^{-1}Q^{-1} \\ &= (\varphi^*Q)R(\varphi)Q^{-1}. \end{aligned} \qquad \ldots(1)$$

Conversely, if R and R' are related as (1), then, for non-degenerate representative matrix functions F and F' with respect to R and R', respectively,

$$\begin{aligned} (\varphi^*F')^{-1} &= R'(\varphi) = (\varphi^*Q)R(\varphi)Q^{-1} \\ &= (\varphi^*Q)(\varphi^*F)F^{-1}Q^{-1} \qquad \text{for all } \varphi \in \tilde{G}. \end{aligned}$$

Hence

$$\varphi^*(F^{-1}Q^{-1}F') = F^{-1}Q^{-1}F' \qquad \text{for all } \varphi \in \tilde{G}.$$

Thus

$$P = F^{-1}Q^{-1}F'$$

is a matrix valued meromorphic function on M with non-zero det(P). This implies that

$$(F') = (QFP) = (FP) = (F)P.$$

In (1), if Q is a constant non-singular matrix, then R' and R are equivalent, $R \approx R'$, in the usual sense.

DEFINITION 3.2.1 Representations R and R' of $\tilde{G}$ of degree r are said to be *weakly equivalent*, $R \sim R'$, if there is an (r x r)-matrix valued holomorphic function Q on $\tilde{M}$ with never vanishing det(Q) such that

$$R'(\varphi) = (\varphi^*Q)R(\varphi)Q^{-1} \qquad \text{for all } \varphi \in \tilde{G}.$$

It is easy to see that this is an equivalence relation. The weak equivalence class is denoted by $R/\sim$.

As an analogy to that in Definition 2.3.1,

DEFINITION 3.2.2 An (r x r)-matrix valued meromorphic function F on $\tilde{M}$ with non-zero det(F) is called a D-*multiplicative (r x r)-matrix function* if, for every $\varphi \in \tilde{G}$, $(\varphi^*F)F^{-1}$ is a constant function.

In this case, if we write $R(\varphi) = (\varphi^*F)F^{-1}$, then

$$R : \tilde{G} \longrightarrow GL(r,\mathbb{C})$$

is a representation of $\tilde{G}$ and F is a non-degenerate representative matrix function. In fact, for $\varphi,\sigma \in \tilde{G}$,

$$R(\varphi\sigma) = (\varphi\sigma)^*(F)F^{-1} = \sigma^*(\varphi^*F)F^{-1} = \sigma^*(R(\varphi)F)F^{-1}$$

$$= R(\varphi)\sigma^*(F)F^{-1} = R(\varphi)R(\varphi).$$

The set of all D-multiplicative (r x r)-matrix functions is denoted by $M(M,D,r)$.

D-multiplicative (r x r)-matrix functions F and F' are said to be *equivalent*, $F \approx F'$ (resp. *weakly equivalent*, $F \sim F'$), if there is an (r x r)-matrix valued meromorphic function P on M with non-zero det(P), and a constant non-singular (r x r)-matrix C (resp. an (r x r)-matrix valued holomorphic function Q on $\tilde{M}$ with never vanishing det(Q)) such that

$$F' = CFP \qquad (\text{resp. } F' = QFP).$$

The (resp. weak) equivalence class is denoted by $F/\approx$ (resp. $F/\sim$).

We denote by $\mathrm{Rep}(\tilde{G},r)$ and $FV(M,D,r)$ the set of all representations of $\tilde{G}$ of degree r and the set of all (isomorphism classes of) flat D-vector bundles of rank r, respectively.

We conclude

PROPOSITION 3.2.3 (1) There are the following one-to-one correspondences:

$$R/\sim \in \mathrm{Rep}(\tilde{G},r)/\sim \;\longmapsto F/\sim \in \mathcal{M}(M,D,r)/\sim \;\longmapsto V_R = [(F)] \in FV(M,D,r),$$

where F is a representative matrix function with respect to R. (2) There is the following one-to-one correspondence:

$$R/\approx \in \mathrm{Rep}(G,r)/\approx \;\longrightarrow F/\approx \in \mathcal{M}(M,D,r)/\approx.$$

(3) The following diagram is commutative:

$$\begin{array}{ccccc} \mathrm{Rep}(\tilde{G},r)/\approx & \rightarrowtail\!\!\!\!\rightarrow & \mathcal{M}(M,D,r)/\approx & & \\ \downarrow\!\!\!\!\downarrow & & \downarrow\!\!\!\!\downarrow & & \\ \mathrm{Rep}(\tilde{G},r)/\sim & \rightarrowtail\!\!\!\!\rightarrow & \mathcal{M}(M,D,r)/\sim & \rightarrowtail\!\!\!\!\rightarrow & FV(M,D,r) \subset V(M,D,r), \end{array}$$

where the symbol ' $\twoheadrightarrow$' means a surjective map and the symbol '$\rightarrowtail\!\!\!\!\rightarrow$ ' means a bijective map.

Our next task is to characterize $FV(M,D,r)$ *in* $V(M,D,r)$.

First of all, we note that, if $F = (f_{jk})$ is an $(r \times r)$-matrix valued meromorphic function on an open set U of $\tilde{M}$, then the differential

$$dF = (df_{jk})$$

is a meromorphic $(r \times r)$-matrix 1-form on U.

LEMMA 3.2.4 Let F be an $(r \times r)$-matrix valued meromorphic function on $\tilde{M}$ with non-zero $\det(F)$. Then F is a D-multiplicative matrix function if and only if $\eta = F^{-1}dF$ is a meromorphic $(r \times r)$-matrix 1-form *on* M.

PROOF. Suppose that, for every $\varphi \in \tilde{G}$, $(\varphi^*F)F^{-1}$ is constant. Then

$$\begin{aligned} 0 &= d(\varphi^*F)F^{-1} + (\varphi^*F)dF^{-1} \\ &= \varphi^*(dF)F^{-1} - (\varphi^*F)F^{-1}(dF)F^{-1}. \end{aligned}$$

Hence

$$\varphi^*(F^{-1}dF) = F^{-1}dF \quad \text{for all } \varphi \in \tilde{G}.$$

Thus $\eta = F^{-1}dF$ is a meromorphic matrix 1-form on M.

The converse is proved by tracing the argument in reverse. □

Note that $\eta = F^{-1}dF$ in the lemma satisfies the integrability condition:

$$d\eta + \eta \wedge \eta = 0.$$

In fact

$$d\eta + \eta \wedge \eta = dF^{-1} \wedge dF + F^{-1}d(dF) + (F^{-1}dF) \wedge (F^{-1}dF)$$
$$= -F^{-1}(dF)F^{-1} \wedge dF + (F^{-1}dF) \wedge (F^{-1}dF) = 0.$$

In general, let η be a meromorphic $(r \times r)$-matrix 1-form on an open set U of $\tilde{M}$ (or, of M), which is said to be *integrable* if

$$d\eta + \eta \wedge \eta = 0.$$

For an integrable η on U, consider the differential equation

$$dF = F\eta \qquad \ldots(2)$$

where F is an $(r \times r)$-matrix valued unknown function. Let $D_\infty(\eta)$ be the polar divisor of $\eta = (\eta_{jk})$:

$$D_\infty(\eta) = \sum_{j,k=1}^{r} D_\infty(\eta_{jk}).$$

It is then well known that the differential equation (2) has a unique holomorphic solution F with never vanishing $\det(F)$ around any point q of $U - \text{Supp}(D_\infty(\eta))$, having any given initial value $F(q)$ at q. Note that, if C is a constant non-singular $(r \times r)$-matrix, then $F' = CF$ is a solution of (2) with the initial value $CF(q)$ at q.

Take a point $q_0 \in U - \text{Supp}(D_\infty(\eta))$. Let F_0 be the solution of (2) around q_0 such that $F_0(q_0) = 1$ (the identity matrix). We continue F_0 analytically along the curves in $U - \text{Supp}(D_\infty(\eta))$.

Following Tōyama [62], we denote by

$$F_0(q) = F_0(q, q_0, \gamma) = {}_e\int_{q_0,\gamma}^{q} \eta,$$

the analytic continuation of F_0 along the curve γ from q_0 to q in $U - \text{Supp}(D_\infty(\gamma))$. We get then a matrix valued and *multi-valued* holomorphic function F_0 on $U - \text{Supp}(D_\infty(\eta))$ with never vanishing $\det(F_0)$. This global function F_0 is also a solution of the differential equation (2).

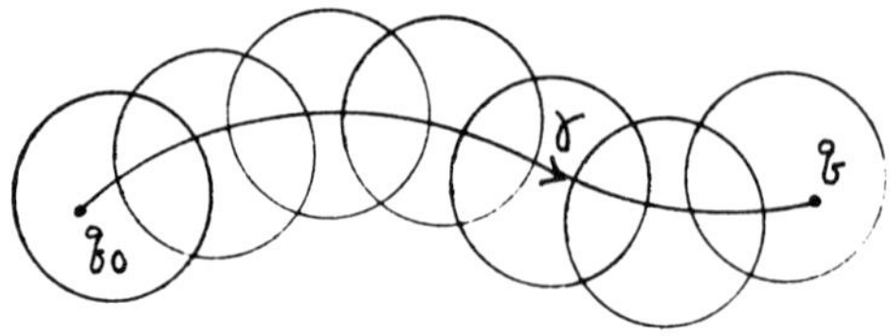

If F is the solution of (2) around q_0 with the initial value $F(q_0)$, then the analytic continuation of F along a curve γ in $U - \mathrm{Supp}(D_\infty(\eta))$ is given by

$$F(q) = F(q, q_0, \gamma) = F(q_0)\ e^{\int_{q_0,\gamma}^{q} \eta}\ .$$

From this remark, we clearly have

LEMMA 3.2.5 Let γ and γ' be curves in $U - \mathrm{Supp}(D_\infty(\eta))$ from q_0 to q_1 and from q_1 to q, respectively. Then

$$e^{\int_{q_0,\gamma\gamma'}^{q} \eta} = e^{\int_{q_0,\gamma}^{q_1} \eta}\ e^{\int_{q_1,\gamma'}^{q} \eta}\ .$$

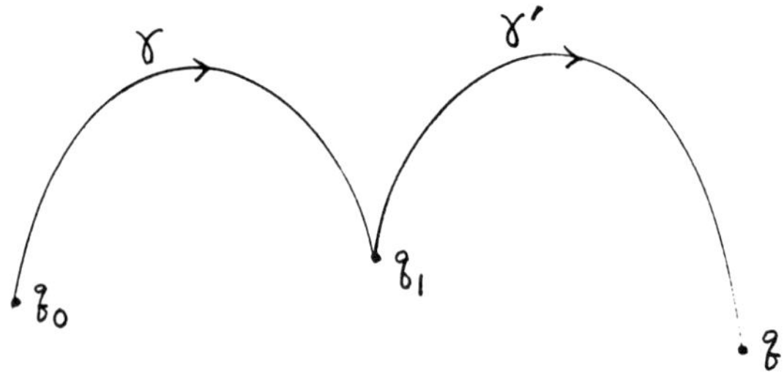

PROPOSITION 3.2.6 (Tōyama). Let η and K be an integrable meromorphic matrix 1-form and a matrix valued meromorphic function with non-zero det(K), respectively, on an open set U of $\tilde{M}$ (or, of M). Then $K\eta K^{-1} - (dK)K^{-1}$ is also integrable and

$$e^{\int_{q_0,\gamma}^{q} (K\eta K^{-1} - (dK)K^{-1})} = K(q_0)\ \left(e^{\int_{q_0,\gamma}^{q} \eta}\right)\ K(q)^{-1},$$

where γ does not pass through $\mathrm{Supp}(D_\infty(\eta))$ nor $\mathrm{Supp}(D_\infty(K\eta K^{-1} - (dK)K^{-1}))$.

PROOF. We write

$$F(q) = \int_{e^{q_o,\gamma}}^{q} \eta \, .$$

Put $L = FK^{-1}$ and put

$$\eta_1 = L^{-1}dL.$$

Then

$$d\eta_1 + \eta_1 \wedge \eta_1 = 0$$

and

$$LK\eta = F\eta = dF = (dL)K + L(dK).$$

Hence

$$\eta = K^{-1}\eta_1 K + K^{-1}dK.$$

Thus

$$\eta_1 = K\eta K^{-1} - (dK)K^{-1}.$$

Since

$$1 = F(q_o) = L(q_o)K(q_o),$$

we have

$$L(q_o) = K(q_o)^{-1}.$$

Now

$$L(q_o) \int_{e^{q_o,\gamma}}^{q} (K\eta K^{-1} - (dK)K^{-1}) = L(q_o) \int_{e^{q_o,\gamma}}^{q} \eta_1$$

$$= L(q) = F(q)K(q)^{-1} = (\int_{e^{q_o,\gamma}}^{q} \eta)K(q)^{-1}.$$

This proves the proposition. □

COROLLARY 3.2.7 Let η be an integrable meromorphic matrix 1-form on an open set U of $\tilde{M}$ (or, of M). Let C be a constant non-singular matrix. Then

$$\int_{e^{q_o,\gamma}}^{q} (C\eta C^{-1}) = C(\int_{e^{q_o,\gamma}}^{q} \eta)C^{-1},$$

where γ does not pass through $\mathrm{Supp}(D_\infty(\eta))$.

For a meromorphic matrix 1-form η on an open set W of M, we sometimes identify $\tilde{\pi}^*\eta$ on $\tilde{\pi}^{-1}(W)$ with η by abuse of notation. Now we prove

<u>THEOREM 3.2.8</u> <u>(Weil)</u> A matrix D-divisor $\hat{E} = \{F_\alpha\}_{\alpha\in A} \cup \{F_\nu\}_{\nu\in N}$ (cf. section 3.1) can be written as $\hat{E} = (F)$ for a D-multiplicative matrix function F if and only if there is an integrable meromorphic matrix 1-form η *on* M such that $\hat{E}\eta\hat{E}^{-1} - (d\hat{E})\hat{E}^{-1}$ is holomorphic, that is,

$$F_\alpha\eta F_\alpha^{-1} - (dF_\alpha)F_\alpha^{-1} \quad (\text{resp. } F_\nu\eta F_\nu^{-1} - (dF_\nu)F_\nu^{-1})$$

is holomorphic on W_α (resp. on U_ν) for every $\alpha \in A$ (resp. $\nu \in N$).

<u>PROOF</u>. Suppose first that $\hat{E} = (F)$ for a D-multiplicative matrix function F. Then there is a matrix valued holomorphic function Q_α (resp. Q_ν) on every W_α (resp. U_ν) with never vanishing $\det(Q_\alpha)$ (resp. $\det(Q_\nu)$) such that

$$F = Q_\alpha F_\alpha \quad \text{on } W_\alpha, \quad (\text{resp. } F = Q_\nu F_\nu \quad \text{on } W_\nu).$$

Hence

$$F^{-1}dF \begin{cases} = F_\alpha^{-1}Q_\alpha^{-1}(dQ_\alpha)F_\alpha + F_\alpha^{-1}dF_\alpha & \text{on } W_\alpha \\ = F_\nu^{-1}Q_\nu^{-1}(dQ_\nu)F_\nu + F_\nu^{-1}dF_\nu & \text{on } U_\nu. \end{cases}$$

Put $\eta = F^{-1}dF$. By Lemma 3.2.4, η is an integrable meromorphic matrix 1-form on M. Moreover,

$$Q_\alpha^{-1}dQ_\alpha = F_\alpha\eta F_\alpha^{-1} - (dF_\alpha)F_\alpha^{-1} \quad \text{on } W_\alpha$$

and

$$Q_\nu^{-1}dQ_\nu = F_\nu\eta F_\nu^{-1} - (dF_\nu)F_\nu^{-1} \quad \text{on } U_\nu$$

are holomorphic matrix 1-forms.

Conversely, suppose that there is an integrable meromorphic matrix 1-form η on M such that

$$\omega_\alpha = F_\alpha\eta F_\alpha^{-1} - (dF_\alpha)F_\alpha^{-1} \quad \text{on} \quad W_\alpha$$

and

$$\omega_\nu = F_\nu\eta F_\nu^{-1} - (dF_\nu)F_\nu^{-1} \quad \text{on} \quad U_\nu$$

are holomorphic. By Lemma 3.2.6, ω_α and ω_ν are integrable. By applying Proposition 3.2.6, we have

$$\int_{e^{q_\alpha,\gamma_\alpha}}^{q} \omega_\alpha = F_\alpha(q_\alpha)\left(\int_{e^{q_\alpha,\gamma_\alpha}}^{q} \eta\right)F_\alpha(q)^{-1} \qquad \text{on } W_\alpha$$

and
$$\int_{e^{q_\nu,\gamma_\nu}}^{q} \omega_\nu = F_\nu(q_\nu)\left(\int_{e^{q_\nu,\gamma_\nu}}^{q} \eta\right)F_\nu(q)^{-1} \qquad \text{on } U_\nu,$$

where $q_\alpha \in W_\alpha - \text{Supp}(D_\infty(\eta))$ and $q_\nu \in U_\nu - \text{Supp}(D_\infty(\eta))$ are fixed points.
Hence

$$\int_{e^{q_\alpha,\gamma_\alpha}}^{q} \eta = F_\alpha(q_\alpha)^{-1}\left(\int_{e^{q_\alpha,\gamma_\alpha}}^{q} \omega_\alpha\right)F_\alpha(q) \qquad \text{on } W_\alpha$$

and

$$\int_{e^{q_\nu,\gamma_\nu}}^{q} \eta = F_\nu(q_\nu)^{-1}\left(\int_{e^{q_\nu,\gamma_\nu}}^{q} \omega_\nu\right)F_\nu(q) \qquad \text{on } U_\nu.$$

These relations show that the matrix valued and multi-valued holomorphic function

$$F(q) = \int_{e^{q_o,\gamma}}^{q} \eta$$

(q_o is a fixed point of $\tilde{M} - \text{Supp}(D_\omega(\eta))$) on $\tilde{M} - \text{Supp}(D_\infty(\eta))$ can be extended to a multi-valued meromorphic function F on $\tilde{M}$, whose set of poles is contained in the union of the sets of poles of F_α's and F_ν's.

These relations also show that the multi-values of F are *not* caused by continuing analytically around the set of poles of η. On the other hand, $\tilde{M}$ is simply connected. Hence F is in fact single-valued on $\tilde{M}$. Moreover, F(q) does not depend on the choice of the curve γ. Hence we may write

$$F(q) = \int_{e^{q_o}}^{q} \eta.$$

By Lemma 3.2.4, F is a D-multiplicative matrix function. We put

$$Q_\alpha(q) = \int_{e^{q_\alpha,\gamma_\alpha}}^{q} \omega_\alpha \qquad \text{on} \quad W_\alpha$$

and

$$Q_\nu(q) = \int_{e^{q_\nu,\gamma_\nu}}^{q} \omega_\nu \qquad \text{on} \quad U_\nu.$$

Then

$$dQ_\alpha = Q_\alpha\omega_\alpha \quad \text{and} \quad dQ_\nu = Q_\nu\omega_\nu .$$

Moreover we have

$$(Q_\alpha F_\alpha)^{-1} d(Q_\alpha f_\alpha) = F_\alpha^{-1} Q_\alpha^{-1} (dQ_\alpha) F_\alpha + F_\alpha^{-1} Q_\alpha^{-1} Q_\alpha dF_\alpha$$
$$= F_\alpha^{-1} \omega_\alpha F_\alpha + F_\alpha^{-1} dF_\alpha = \eta.$$

In a similar way, we have

$$(Q_\nu F_\nu)^{-1} d(Q_\nu F_\nu) = \eta.$$

Hence there are constant non-singular matrices C_α and C_ν such that

$$Q_\alpha F_\alpha = C_\alpha F \qquad \text{on} \qquad W_\alpha$$

and

$$Q_\nu F_\nu = C_\nu F \qquad \text{on} \qquad U_\nu.$$

Thus

$$\hat{E} = (F). \quad \square$$

Let $\hat{E} = \{F_\alpha\}_{\alpha \in A} \cup \{F_\nu\}_{\nu \in N}$ be a matrix D-divisor. Then

$$[\hat{E}] = \{F_{\alpha\beta}\} \cup \{F_{\alpha\nu}\} \cup \{F_{\nu\alpha}\} \cup \{F_{\nu\mu}\},$$

where $F_{\alpha\beta} = F_\alpha F_\beta^{-1}$, etc. Let S_ν be a matrix valued holomorphic function on U with never vanishing $\det(S_\nu)$ such that

$$\sigma_\nu^* F_\nu = S_\nu F_\nu \qquad \text{(see section 3.1).}$$

Under these notations,

<u>THEOREM 3.2.9</u> <u>(Atiyah)</u> $[\hat{E}]$ is a flat D-vector bundle if and only if there is an integrable holomorphic matrix 1-form ω_α (resp. ω_ν) on W_α (resp. on U_ν) for every $\alpha \in A$ (resp. $\nu \in N$) such that (1)

$$\omega_\alpha = F_{\alpha\beta}\omega_\beta F_{\alpha\beta}^{-1} - (dF_{\alpha\beta})F_{\alpha\beta}^{-1} \quad \text{on} \quad W_\alpha \cap W_\beta \quad \text{for} \quad \alpha,\beta \in A,$$
$$\omega_\alpha = F_{\alpha\nu}\omega_\nu F_{\alpha\nu}^{-1} - (dF_{\alpha\nu})F_{\alpha\nu}^{-1} \quad \text{on} \quad W_\alpha \cap W_\nu \quad \text{for} \quad \alpha \in A \text{ and } \nu \in N,$$
$$\omega_\nu = F_{\nu\mu}\omega_\mu F_{\nu\mu}^{-1} - (dF_{\nu\mu})F_{\nu\mu}^{-1} \quad \text{on} \quad U_\nu \cap U_\mu \quad \text{for} \quad \nu,\mu \in N,$$

and (2)

$$\sigma_\nu^* \omega_\nu = S_\nu \omega_\nu S_\nu^{-1} - (dS_\nu)S_\nu^{-1} \quad \text{on} \quad U_\nu \quad \text{for} \quad \nu \in N.$$

<u>PROOF</u>. By conditions (1) and (2), we can easily show that

$$\eta = F_\alpha^{-1}\omega_\alpha F_\alpha + F_\alpha^{-1} dF_\alpha = F_\nu^{-1}\omega_\nu F_\nu + F_\nu^{-1} dF_\nu$$

is an integrable meromorphic matrix 1-form on M. Since $\hat{E}\eta\hat{E}^{-1} - (d\hat{E})\hat{E}^{-1}$ is holomorphic, $[\hat{E}]$ is a flat D-vector bundle, by Theorem 3.2.8.

The converse also follows easily from Theorem 3.2.8. □

REMARK 3.2.10 (1) Theorem 3.2.9 can be considered as a generalization of Theorem 2.3.6. (2) If $\bar{B}$ is empty, then Theorem 3.2.9 says that a holomorphic vector bundle V on M is flat if and only if V has an integrable (i.e. flat) holomorphic connection. See Atiyah [2].

For an (r x r)-matrix D-divisor $\hat{E} = \{F_\alpha\} \cup \{F_\nu\}$, the D-vector bundle $[\hat{E}]$ associated with $\hat{E}$ is said to be *decomposable* if there is an $(r_1 x r_1)$-matrix (resp. an $(r_2 x r_2)$-matrix) D-divisor $E_1 = \{F_{1\alpha}\} \cup \{F_{1\nu}\}$ (resp. $\hat{E}_2 = \{F_{2\alpha}\} \cup \{F_{2\nu}\}$) with $r_1+r_2 = r$ such that

$$[\hat{E}] = [\hat{E}_1 \oplus \hat{E}_2],$$

where

$$\hat{E}_1 \oplus \hat{E}_2 = \left\{\begin{pmatrix} F_{1\alpha} & 0 \\ 0 & F_{2\alpha} \end{pmatrix}\right\} \cup \left\{\begin{pmatrix} F_{1\nu} & 0 \\ 0 & F_{2\nu} \end{pmatrix}\right\} .$$

In this case, we write

$$[\hat{E}] = [\hat{E}_1] \oplus [\hat{E}_2].$$

(This is well defined.) Otherwise, $[\hat{E}]$ is said to be *indecomposable*.

An *indecomposable decomposition of* $[\hat{E}]$ is, by definition,

$$[\hat{E}] = [\hat{E}_1] \oplus \dots \oplus [\hat{E}_m],$$

where every $[\hat{E}_j]$ is indecomposable. Every $[\hat{E}_j]$ is called an *indecomposable component of* $[\hat{E}]$.

If every $\hat{E}_j$ can be written as

$$\hat{E}_j = (\hat{F}_j)$$

for a D-multiplicative matrix function F_j, then

$$c(\det[\hat{E}_j]) = 0 \quad \text{for} \quad 1 \leq j \leq m,$$

by Theorem 2.3.6.

The following theorem is due to Weil [64]. See also Gunning [25].

THEOREM 3.2.11 (Weil) Let $\bar{M} = M$ be a compact Riemann surface. For a D-vector bundle $[\hat{E}]$ on M, let $[\hat{E}] = [\hat{E}_1] \oplus \dots \oplus [\hat{E}_m]$ be an indecomposable decomposition of $[\hat{E}]$. Then $[\hat{E}]$ is flat if and only if $c(\det[\hat{E}_j]) = 0$ for

$1 \leq j \leq m$.

But, if $\dim \bar{M} \geq 2$, then this theorem does not necessarily hold. In fact, Oda [48] gave examples of indecomposable vector bundles V on an abelian surface $\bar{M} = M$ such that $c(\det V) = 0$ and that V has no flat connection.

The conditions of Theorem 3.2.9 seem difficult to handle. It is desirable to replace them by other conditions which are easy to handle, as those in Theorem 3.2.11.

Our final task in this section is to describe the fibres of the surjective map

$$R/\approx \in \mathrm{Rep}(\tilde{G},r)/\approx \longmapsto R/\sim \in \mathrm{Rep}(\tilde{G},r)/\sim,$$

(see Proposition 3.2.3).

<u>DEFINITION 3.2.12</u> Let $R : \tilde{G} \longrightarrow GL(r,\mathbb{C})$ be a representation of $\tilde{G}$. An integrable holomorphic $(r \times r)$-matrix 1-form ω (resp. an $(r \times r)$-matrix valued holomorphic function K with never vanishing $\det(K)$) on $\tilde{M}$ is called a *kernel differential* (resp. *function*) *of* R if

$$\varphi^*\omega = R(\varphi)\omega R(\varphi)^{-1} \qquad \text{for all} \quad \varphi \in \tilde{G},$$

$$(\text{resp. } \varphi^*K = R(\varphi)KR(\varphi)^{-1} \qquad \text{for all} \quad \varphi \in \tilde{G}).$$

Note that, if $\omega = 0$ (resp. $K = 1$) identically, then ω (resp. K) is a kernel differential (resp. function) of R.

Note that, for an integrable holomorphic matrix 1-form ω on $\tilde{M}$, $e^{\int_{q_0,\gamma}^{q}} \omega$ does not depend on the choice of the path γ, because $\tilde{M}$ is simply connected. We may thus denote it by

$$e^{\int_{q_0}^{q}} \omega .$$

This is a matrix valued holomorphic function on $\tilde{M}$ with the never vanishing determinant. Conversely, by such a function Q on $\tilde{M}$, we have an integrable holomorphic matrix 1-form $\omega = Q^{-1}dQ$.

Now, suppose that a representation

$$R : \tilde{G} \longrightarrow GL(r,\mathbb{C})$$

is given. Let R' be another representation of $\tilde{G}$ such that $R' \sim R$ (weakly equivalent). That is, there is a matrix valued holomorphic function Q with never vanishing $\det(Q)$ such that

$$R'(\varphi) = (\varphi^*Q)R(\varphi)Q^{-1} \quad \text{for all } \varphi \in \tilde{G}.$$

Then we have

$$\varphi^*(Q^{-1}dQ) = R(\varphi)(Q^{-1}dQ)R(\varphi)^{-1} \quad \text{for all } \varphi \in \tilde{G}.$$

Hence $Q^{-1}dQ$ is a kernel differential of R.

Conversely, let ω be a kernel differential of R. Put

$$Q(q) = e^{\int_{q_0}^{q} \omega}.$$

Then, for any fixed $\varphi \in \tilde{G}$,

$$R'(\varphi) = (\varphi^*Q)R(\varphi)Q^{-1}$$

is a constant non-singular matrix. In fact,

$$\begin{aligned} d((\varphi^*Q)R(\varphi)Q^{-1}) &= (\varphi^*dQ)R(\varphi)Q^{-1} - (\varphi^*Q)R(\varphi)Q^{-1}(dQ)Q^{-1} \\ &= (\varphi^*Q) \;\{ \varphi^*(Q^{-1}dQ)R(\varphi) - R(\varphi)(Q^{-1}dQ) \}\, Q^{-1} \\ &= (\varphi^*Q) \;\{ \varphi^*\omega - R(\varphi)\omega R(\varphi)^{-1} \}\, R(\varphi)Q^{-1} = 0. \end{aligned}$$

Moreover, for $\varphi,\ \sigma \in \tilde{G}$,

$$\begin{aligned} R'(\varphi)R'(\sigma) &= (\varphi^*Q)R(\varphi)Q^{-1}(\sigma^*Q)R(\sigma)Q^{-1} \\ &= \sigma^*((\varphi^*Q)R(\varphi)Q^{-1})(\sigma^*Q)R(\sigma)Q^{-1} \\ &= ((\varphi\sigma)^*Q)R(\varphi)(\sigma^*Q^{-1})(\sigma^*Q)R(\sigma)Q^{-1} \\ &= ((\varphi\sigma)^*Q)R(\varphi)R(\sigma)Q^{1} = R'(\varphi\sigma). \end{aligned}$$

Hence R' is a representation of $\tilde{G}$, which is weakly equivalent to R.

Let C be a constant non-singular matrix. If the above function Q is replaced by CQ, then

$$(\varphi^*(CQ))R(\varphi)(CQ)^{-1} = CR'(\varphi)C^{-1} \quad \text{for } \varphi \in \tilde{G}.$$

Hence we get a representation $CR'C^{-1}$ which is equivalent to R'.

Thus we have a surjective map

$$\omega \in \{\text{kernel differentials of } R\} \longmapsto R'/\approx\ \in \{R'/\approx \mid R' \sim R\},$$

where $R'(\varphi) = (\varphi^*Q)R(\varphi)Q^{-1}$.

Suppose that ω and ω' correspond to the same $R'/\approx$. Then there is a constant non-singaulr matrix C such that

$$(\varphi^*Q')R(\varphi)Q'^{-1} = C(\varphi^*Q)R(\varphi)Q^{-1}C^{-1} \quad \text{for} \quad \varphi \in G,$$

where

$$Q'(q) = e\int_{q_o}^{q} \omega'.$$

Then

$$\varphi^*((CQ)^{-1}Q') = R(\varphi)((CQ)^{-1}Q')R(\varphi)^{-1} \quad \text{for} \quad \varphi \in \tilde{G}.$$

Thus

$$K = (CQ)^{-1}Q'$$

is a kernel function of R. We have then

$$\omega' = Q'^{-1}dQ' = (CQK)^{-1}d(CQK) = K^{-1}Q^{-1}((dQ)K + QdK)$$
$$= K^{-1}\omega K + K^{-1}dK.$$

Conversely, for any kernel differential ω of R and for any kernel function K or R,

$$\omega' = K^{-1}\omega K + K^{-1}dK$$

is a kernel differential of R. Moreover, ω and ω' correspond to the same equivalence class $R'/\approx$, as is easily seen.

We say that two kernel differentials ω and ω' of R are said to be *equivalent*, $\omega \approx \omega'$, if there is a kernel function K or R such that

$$\omega' = K^{-1}\omega K + K^{-1}dK.$$

This is an equivalence relation, as is easily seen.

We conclude

PROPOSITION 3.2.13 Let R be a fixed representation of $\tilde{G}$. Then there is a bijective map

$$\{\text{kernel differentials of } R\} / \approx \rightarrowtail\!\!\!\rightarrow \{R'/\approx \mid R' \sim R\}.$$

The following lemma is clear from the definition.

LEMMA 3.2.14 For any integrable meromorphic matrix 1-form η on $\tilde{M}$ and for any $\varphi \in \tilde{G}$,

$$e\int_{q_o,\gamma}^{q} \varphi^*\eta = e\int_{\varphi(q_o),\varphi(\gamma)}^{\varphi(q)} \eta,$$

where γ is a curve in $\tilde{M}$ from q_o to q not passing through $\mathrm{Supp}(D_\infty(\varphi^*\eta))$.

THEOREM 3.2.15 (Tōyama) Let R and R' be representations of $\tilde{G}$. R' is weakly equivalent to R if and only if there is a kernel differential ω of R and a constant non-singular matrix C such that

$$R'(\varphi) = C(e^{\int_{q_0}^{\varphi(q_0)}} \omega)R(\varphi)C^{-1} \quad \text{for all } \varphi \in \tilde{G}.$$

(q_0 is a fixed point of $\tilde{M}$.)

PROOF. If R' is weakly equivalent to R, then there is a matrix valued holomorphic function Q with never vanishing det(Q) such that

$$R'(\varphi) = (\varphi^*Q)R(\varphi)Q^{-1} \quad \text{for } \varphi \in \tilde{G}.$$

We may write

$$Q(q) = C\, e^{\int_{q_0}^{q}} \omega,$$

where $C = Q(q_0)$ and $\omega = Q^{-1}dQ$. As was shown above, ω is a kernel differential of R.

Now, by Corollary 3.2.7, Lemmas 3.2.14 and 3.2.5, we have

$$\begin{aligned} R'(\varphi) &= C(\varphi^* e^{\int_{q_0}^{q}} \omega)R(\varphi)(e^{\int_{q_0}^{q}} \omega)^{-1}C^{-1} \\ &= C(e^{\int_{q_0}^{\varphi(q)}} \omega)R(\varphi)(e^{\int_{q}^{q_0}} \omega)R(\varphi)^{-1}R(\varphi)C^{-1} \\ &= C(e^{\int_{q_0}^{\varphi(q)}} \omega)(e^{\int_{q}^{q_0}} R(\varphi)\omega R(\varphi)^{-1})R(\varphi)C^{-1} \\ &= C(e^{\int_{q_0}^{\varphi(q)}} \omega)(e^{\int_{q}^{q_0}} \varphi^*\omega)R(\varphi)C^{-1} \\ &= C(e^{\int_{q_0}^{\varphi(q)}} \omega)(e^{\int_{\varphi(q)}^{\varphi(q_0)}} \omega)R(\varphi)C^{-1} \\ &= C(e^{\int_{q_0}^{\varphi(q_0)}} \omega)R(\varphi)C^{-1} \quad \text{for } \varphi \in \tilde{G}. \end{aligned}$$

Conversely, given ω and C, we define R' by

$$R'(\varphi) = C(e^{\int_{q_0}^{\varphi(q_0)}} \omega)R(\varphi)C^{-1} \quad \text{for } \varphi \in \tilde{G}.$$

By tracing the above argument in reverse, we have

$$R'(\varphi) = (\varphi^*Q)R(\varphi)Q^{-1},$$

where

$$Q(q) = C\, e^{\int_{q_0}^{q} \omega}.$$

Hence R' is a representation of $\tilde{G}$ which is weakly equivalent to R. □

3.3 Unitary representations

We follow the notations and the assumptions in section 3.1.

The weak equivalence of representations (cf. section 3.2) is important, but is not easy to handle. As for unitary representations, however,

THEOREM 3.3.1 (Weil-Tōyama). Unitary representations R and R' of $\tilde{G}$ are weakly equivalent if and only if they are equivalent.

PROOF. For a matrix $A = (a_{jk})$, we define the norm $\|A\|$ by

$$\|A\| = \sqrt{\sum_{j,k} |a_{jk}|^2} = \sqrt{\mathrm{Trace}({}^t\bar{A}A)},$$

where ${}^t\bar{A}$ is the transpose of the complex conjugate of A. Let

$$R : \tilde{G} \longrightarrow U(r)$$

and

$$R' : \tilde{G} \longrightarrow U(r)$$

be unitary representations. ($U(r)$ ($\subset GL(r,\mathbb{C})$) is the unitary group of degree r.) Suppose that R and R' are weakly equivalent, that is, there is a matrix valued holomorphic function Q on $\tilde{M}$ with never vanishing det(Q) such that

$$R'(\varphi) = (\varphi^*Q)R(\varphi)Q^{-1} \quad \text{for all } \varphi \in \tilde{G}.$$

Then we have

$${}^t\overline{R(Q)}\varphi^*\,{}^t\bar{Q} = {}^t\bar{Q}\,{}^t\overline{R'(\varphi)}.$$

Hence

$$R(\varphi)^{-1}\varphi^*\,{}^t\bar{Q} = {}^t\overline{R(\varphi)}\varphi^*\,{}^t\bar{Q} = {}^t\bar{Q}\,{}^t\overline{R'(\varphi)} = {}^t\bar{Q}R'(\varphi)^{-1}$$
$$= {}^t\bar{Q}QR(\varphi)^{-1}\varphi^*Q^{-1}.$$

Hence

$$\varphi^*({}^t\overline{Q}Q) = R(\varphi)({}^t\overline{Q}Q)R(\varphi)^{-1}.$$

By taking the trace, we have

$$\varphi^*\|Q\|^2 = \|Q\|^2 \quad \text{for all } \varphi \in \tilde{G}.$$

This means that $\|Q\|^2$ is a continuous function *on* M. Since $\|Q\|^2$ can be written as

$$\|Q\|^2 = \sum_{j,k} |h_{jk}|^2,$$

where $Q = (h_{jk})$, $\|Q\|^2$ is a pluri-subharmonic function on M, (see, for example, Grauert [20]). Since

$$\mathrm{codim}(\bar{M} - M) = \mathrm{codim}(\mathrm{Sing}\bar{B}) \geq 2,$$

$\|Q\|^2$ can be extended to a pluri-subharmonic function on the compact $\bar{M}$. Hence $\|Q\|^2$ must be constant.

Hence, for any vector $\xi = (\xi_1,\ldots,\xi_n) \in \mathbb{C}^n$,

$$0 = \sum_{\alpha,\beta} \frac{\partial^2\|Q\|}{\partial z_\alpha \partial \bar{z}_\beta} \xi_\alpha \bar{\xi}_\beta = \sum_{j,k} \left| \sum_\alpha \frac{\partial h_{jk}}{\partial z_\alpha} \xi_\alpha \right|^2 ,$$

where $(z_1,\ldots,z_n)$ is a local coordinate system in $\tilde{M}$.
Hence

$$\sum_\alpha \frac{\partial h_{jk}}{\partial z_\alpha} \xi_\alpha = 0 \quad \text{for all } \xi \in \mathbb{C}^n \text{ and all } j,k.$$

Hence

$$\frac{\partial h_{jk}}{\partial z_\alpha} = 0 \quad \text{for all } j,k \text{ and } \alpha.$$

Thus Q = C must be a constant non-singular matrix. The relation

$$R'(\varphi) = CR(\varphi)C^{-1} \quad \text{for all } \varphi \in \tilde{G}$$

means that R and R' are equivalent.

The converse is trivial. □

As an analogy to Schur's lemma,

<u>THEOREM 3.3.2</u> <u>(Tōyama)</u> Let $R : \tilde{G} \longrightarrow GL(r,\mathbb{C})$ and $R' : \tilde{G} \longrightarrow GL(r',\mathbb{C})$ be representations of $\tilde{G}$ both of which are weakly equivalent to irreducible unitary representations. Suppose that there is an $(r' \times r)$-matrix valued holomorphic function Z on $\tilde{M}$ such that

$$R'(\varphi)Z = (\varphi^*Z)R(\varphi) \quad \text{for all } \varphi \in \tilde{G}.$$

Then, either $Z = 0$ (identically) or (1) $r = r'$, (2) $\det(Z)$ is never vanishing and (3) $R \sim R'$ (weakly equivalent).

PROOF. By assumption, there is an $(r \times r)$-matrix valued (resp. $(r' \times r)$-matrix valued) holomorphic function Q (resp. Q') with never vanishing $\det(Q)$ (resp. $\det(Q')$) and an irreducible unitary representation U (resp. U') of $\tilde{G}$ such that

$$U(\varphi) = (\varphi^*Q)R(\varphi)Q^{-1} \qquad \text{for all } \varphi \in \tilde{G},$$

$$(\text{resp. } U'(\varphi) = (\varphi^*Q')R'(\varphi)Q'^{-1} \qquad \text{for all } \varphi \in \tilde{G}).$$

Then

$$U'(\varphi)(Q'ZQ^{-1}) = \varphi^*(Q'ZQ^{-1})U(\varphi) \quad \text{for all } \varphi \in \tilde{G}.$$

By the same argument as that for the proof of Theorem 3.3.1,

$$C = Q'ZQ^{-1}$$

is a constant $(r' \times r)$-matrix. By Schur's lemma, either $C = 0$ or $r = r'$ and $\det(C) \neq 0$. Hence, either $Z = 0$ identically or (1) $r = r'$, (2) $\det(Z) = \det(Q'^{-1}CQ)$ is never vanishing and (3) $R \sim U \approx U' \sim R'$. □

By Theorem 3.2.15,

PROPOSITION 3.3.3 A representation R of $\tilde{G}$ is weakly equivalent to a unitary representation if and only if there is a kernel differential ω of R such that the representation

$$\varphi \in \tilde{G} \longmapsto \left(e^{\int_{q_0}^{\varphi(q_0)} \omega}\right)R(\varphi) \in GL(r,\mathbb{C})$$

is equivalent to a unitary representation.

It should be noted that, in general, not every representation is weakly equivalent to a unitary representation. The following example is due to Professor Iwasawa.

EXAMPLE 3.3.4 (Iwasawa) Let $\bar{M} = M$ be a compact Riemann surface of genus g (≥ 1) and $\bar{B} = B$ be empty. As in section 1.2, let $\alpha_1,\ldots,\alpha_g,\beta_1,\ldots,\beta_g$ be generators of $\pi_1(M,p_0)$ with the unique relation

$$\alpha_1\beta_1\alpha_1^{-1}\beta_1^{-1}\cdots\alpha_g\beta_g\alpha_g^{-1}\beta_g^{-1} = 1.$$

Take a non-zero complex number b. Consider the representation

$$R : \tilde{G} = \pi_1(M,p_o) \longrightarrow GL(2,\mathbb{C})$$

defined by

$$R(\alpha_1) = \begin{pmatrix} 1 & b \\ 0 & 1 \end{pmatrix}$$

$$R(\alpha_j) = \begin{pmatrix} 1 & 0 \\ 0 & 1 \end{pmatrix} \qquad \text{for } 2 \leq j \leq g$$

and

$$R(\beta_k) = \begin{pmatrix} 1 & 0 \\ 0 & 1 \end{pmatrix} \qquad \text{for } 1 \leq k \leq g.$$

We show that R is *not* weakly equivalent to any unitary representation. Suppose the converse. Then there is a unitary representation $R' : \tilde{G} \longrightarrow U(2)$ and a (2 x 2)-matrix valued holomorphic function Q on $\tilde{M}$ (= the universal covering space of M) with never vanishing det(Q) such that

$$R'(\varphi) = (\varphi^*Q)R(\varphi)Q^{-1} \qquad \text{for all } \varphi \in \tilde{G}.$$

Put

$$Q = \begin{pmatrix} h_{11} & h_{12} \\ h_{21} & h_{22} \end{pmatrix}$$

where h_{jk} are holomorphic functions on M.

Since, for any $\varphi \in \tilde{G}$, $R(\varphi)$ can be written as

$$R(\varphi) = \begin{pmatrix} 1 & b_\varphi \\ 0 & 1 \end{pmatrix} \quad (b_\varphi \in \mathbb{C}),$$

we have

$$R'(\varphi)Q = (\varphi^*Q)R(\varphi) = \begin{pmatrix} \varphi^*h_{11} & b_\varphi\varphi^*h_{11}+\varphi^*h_{12} \\ \varphi^*h_{21} & b_\varphi\varphi^*h_{21}+\varphi^*h_{22} \end{pmatrix}.$$

Writing the first columns, we have

$$R'(\varphi)\begin{pmatrix} h_{11} \\ h_{21} \end{pmatrix} = \varphi^*\begin{pmatrix} h_{11} \\ h_{21} \end{pmatrix}. \ 1 \qquad \text{for all } \varphi \in \tilde{G}.$$

This relation and Theorem 3.3.2 show that the representation R' must be reducible. Hence there is a constant non-singular (2 x 2)-matrix C such that

$$R'(\varphi) = C\begin{pmatrix} a_1(\varphi) & 0 \\ 0 & a_2(\varphi) \end{pmatrix} C^{-1} \qquad \text{for all } \varphi \in \tilde{G}.$$

where

$$|a_1(\varphi)| = |a_2(\varphi)| = 1.$$

By replacing Q by $C^{-1}Q$, we may assume that

$$\begin{pmatrix} a_1(\varphi) & 0 \\ 0 & a_2(\varphi) \end{pmatrix} Q = (\varphi^*Q)R(\varphi) \qquad \text{for all } \varphi \in \tilde{G}.$$

Hence, for any $\varphi \in \tilde{G}$,

$$\varphi^*h_{11} = a_1(\varphi)h_{11}, \quad \varphi^*h_{21} = a_2(\varphi)h_{21},$$

$$b_\varphi \varphi^*h_{11} + \varphi^*h_{12} = a_1(\varphi)h_{12},$$

and

$$b_\varphi \varphi^*h_{21} + \varphi^*h_{22} = a_2(\varphi)h_{22}.$$

The first two equations imply that

$$|\varphi^*h_{11}|^2 = |h_{11}|^2$$

and

$$|\varphi^*h_{21}|^2 = |h_{21}|^2$$

are continuous subharmonic functions on M, hence are constant. Thus

$$h_{11} = c_{11} \quad \text{and} \quad h_{21} = c_{21}$$

are themselves constant. Moreover

$$a_1(\varphi) = a_2(\varphi) = 1 \qquad \text{for all } \varphi \in \tilde{G}.$$

By the above last two equations,

$$\varphi^* h_{12} = h_{12} - b_\varphi c_{11}$$
$$\varphi^* h_{22} = h_{22} - b_\varphi c_{21} \qquad \text{for all } \varphi \in \tilde{G}.$$

This means that h_{12} and h_{22} are additive functions, (see section 2.2). Since h_{12} and h_{22} are holomorphic

$$\omega_{12} = \varphi^* dh_{12} = dh_{12}$$

and

$$\omega_{22} = \varphi^* dh_{22} = dy_{22}$$

are holomorphic differentials on M. The argument in section 2.2 shows that

$$-b_\varphi c_{11} = \int_\varphi \omega_{12}$$
$$-b_\varphi c_{21} = \int_\varphi \omega_{22} \qquad \text{for all } \varphi \in \tilde{G}.$$

(Here φ is identified with a closed path in M.)

In particular, since $b_{\beta_j} = 0$,

$$\int_{\beta_j} \omega_{12} = \int_{\beta_j} \omega_{22} = 0 \qquad \text{for all } 1 \leq j \leq g.$$

This implies that

$$\omega_{12} = \omega_{22} = 0,$$

(see, for example, Namba [44, Corollary 5.2.6]). Hence h_{12} and h_{22} are constant. Thus

$$b_\varphi c_{11} = b_\varphi c_{21} = 0 \qquad \text{for all } \varphi \in \tilde{G}.$$

But

$$b_{\alpha_1} = b \neq 0.$$

Hence

$$c_{11} = c_{21} = 0.$$

This contradicts the assumption that det(Q) is never vanishing. Hence R cannot be weakly equivalent to a unitary representation.

Unitary representations are important in the present study, (cf. Chapter 2). As in section 2.3,

DEFINITION 3.3.5 A matrix valued meromorphic function F on $\tilde{M}$ is called a *strict D-multiplicative matrix function* if F is a D-multipicative matrix function such that

$$R(\varphi) = (\varphi^* F)F^{-1}$$

is a unitary matrix for every $\varphi \in \tilde{G}$. The flat D-vector bundle V_R associated with a unitary representation R is called a *unitary flat D-vector bundle.*

The set of all strict D-multiplicative (r x r)-matrix functions is denoted by $M_s(M,D,r)$. The set of all (isomorphism classes of) unitary flat D-vector bundles of rank r is denoted by UFV(M,D,r). Also, we put

$$\mathrm{Uni}(\tilde{G},r) = \{R \mid R \text{ is a unitary representation of } \tilde{G} \text{ of degree } r\}$$

and

$$\mathrm{Div}_s(M,D,r) = \{(F) \mid F \text{ is a strict D-multiplicative } (r \times r)\text{-matrix function}\}.$$

Then Theorem 3.3.1 and Proposition 3.2.3 say that there are the following bijective correspondences:

$$\mathrm{Uni}(\tilde{G},r)/\approx \;\rightarrowtail\; M_s(M,D,r)/\approx \;\rightarrowtail\; \mathrm{Div}_s(M,D,r)/\sim$$
$$\rightarrowtail\; \mathrm{UFV}(M,D,r).$$

Now, let F be a D-multiplicative matrix function. Put $\eta = F^{-1}dF$. Then η is a meromorphic matrix 1-form on M. F can be written as

$$F(q) = F(q_0)\; e\!\!\int_{q_0}^{q} \eta,$$

where q_0 is a fixed point of $\tilde{M}$ - $\mathrm{Supp}(D_\infty(\eta))$. The representation R corresponding to F is defined by

$$R(\varphi) = (\varphi^* F)F^{-1} \qquad \text{for } \varphi \in \tilde{G}.$$

Hence, by Lemmas 3.2.14 and 3.2.5,

$$R(\varphi) = F(q_0)\Big(e\!\!\int_{q_0}^{\varphi(q)} \eta\Big)\Big(e\!\!\int_{q}^{q_0} \eta\Big)F(q_0)^{-1}$$

$$= F(q_0)\Big(e\!\!\int_{q_0}^{\varphi(q)} \eta\Big)\Big(e\!\!\int_{q}^{q_0} \varphi^*\eta\Big)F(q_0)^{-1}$$

$$= F(q_0)\left(\int_{e\, q_0}^{\varphi(q)} \eta\right)\left(\int_{e\, \varphi(q)}^{\varphi(q_0)} \eta\right)F(q_0)^{-1}$$

$$= F(q_0)\left(\int_{e\, q_0}^{\varphi(q_0)} \eta\right)F(q_0)^{-1}.$$

Thus

$$R_\eta : \varphi \in \tilde{G} \longmapsto \int_{e\, q_0}^{\varphi(q_0)} \eta \in GL(r,\mathbb{C})$$

is a representation of $\tilde{G}$ which is equivalent to R. If we regard φ as a closed path in M which starts from $p_0 = \tilde{\pi}(q_0)$, then $\int_{e\, q_0}^{\varphi(q_0)} \eta$ can be written as $\int_{e\, \varphi} \eta$. Thus we may write

$$R_\eta : \varphi \in \tilde{G} \longmapsto \int_{e\, \varphi} \eta \in GL(r,\mathbb{C}).$$

We call R_η the *period representation of* $\tilde{G}$ *for* η.

By Theorem 3.2.8,

THEOREM 3.3.6 A matrix D-divisor $\hat{F}$ can be written as $\hat{E} = (F)$ for a strict D-multiplicative matrix function F if and only if there is an integrable meromorphic matrix 1-form η on M such that (1) $\hat{E}\eta\hat{E}^{-1} - (d\hat{E})\hat{E}^{-1}$ is holomorphic and (2) the period representation R_η for η is equivalent to a unitary representation of $\tilde{G}$.

REMARK 3.3.7 This theorem is not satisfactory as a characterization of $Div_s(M,D,r)$ in $Div(M,D,r)$. It is desirable to replace condition (2) by a condition which relates to neither $\tilde{M}$ nor $\tilde{G}$, that is, a condition which comes from the inside of M (according to the spirit of class field theory).

For a compact Riemann surface M, Narasimhan-Seshadri [46] and Seshadri [56] gave very deep analysis on the structure of UFV(M,D,r). In particular, in the case in which B is empty, they proved the following beautiful theorem.

THEOREM 3.3.8 (Narasimhan-Seshadri) A holomorphic vector bundle V on a compact Riemann surface M of genus $g \geq 2$ is a unitary flat vector bundle if and only if every indecomposable component of V is of degree 0 and stable.

Here, the degree deg(W) of a holomorphic vector bundle W on M is the degree of the line bundle det(W). W is said to be *stable* if, for every proper subbundle W' ($0 \subsetneq W' \subsetneq W$) of W, the following inequality holds:

$$\frac{\deg(W')}{\mathrm{rank}(W')} < \frac{\deg(W)}{\mathrm{rank}(W)} .$$

For later use, we give here definitions of flat D-*vector bundles and unitary flat* D-*vector bundles which are independent of Condition 1.2.4.* First, note that a D-divisor and a D-vector bundle can be defined independently from Condition 1.2.4, as at the end of section 3.1. A D-vector bundle V of rank r is said to be *flat*, if there is a D-divisor $\hat{E} = \{F_\alpha\} \cup \{F_\nu\}$ with $[\hat{E}] = V$ and an integrable meromorphic matrix 1-form η on M such that $\hat{E}\eta\hat{E}^{-1} - (d\hat{E})\hat{E}^{-1}$ is holomorphic. FV(M,D,r) is, by definition, the set of all flat D-vector bundles of rank r.

Let V, $\hat{E}$ and η be as above. Consider the matrix valued, multi-valued holomorphic function

$$F(p) = e^{\int_{p_0,\gamma}^{p}} \eta$$

on $M - \mathrm{Supp}(D_\infty(\eta))$. ($\gamma$ is a curve from p_0 to p.) Using the same notations as in the proof of Theorem 3.2.8, we have

$$e^{\int_{p_\alpha,\gamma_\alpha}^{p}} \eta = F_\alpha(p_\alpha)^{-1} \left(e^{\int_{p_\alpha,\gamma_\alpha}^{p}} \omega_\alpha\right) F_\alpha(p) \qquad \text{on} \quad W_\alpha,$$

where p_α is a fixed point of W_α, γ_α is a curve in W_α from p_α to p and $\omega_\alpha = F_\alpha \eta F_\alpha^{-1} - (dF_\alpha)F_\alpha^{-1}$ is holomorphic. ($B \cap W_\alpha = \emptyset$.)

This shows that F can be extended meromorphically on M - B, that is, F can be considered as a matrix valued meromorphic function on $(M - B)_0$, where

$$\mu_0 : (M - B)_0 \longrightarrow M - B$$

is the universal covering space of M - B. As before, we can then prove that

$$R_\eta : \gamma \in \pi_1(M\text{-}B,\ p_0) \longmapsto e^{\int_\gamma} \eta \in GL(r,\mathbb{C})$$

is a representation of $\pi_1(M\text{-}B,\ p_0)$, called the *period representation of* $\pi_1(M\text{-}B,\ p_0)$ *for* η. If η can be chosen so that R_η is equivalent to a unitary representation, then $V = [\hat{E}]$ is called a *unitary flat* D-*vector bundle of rank* r. UFV(M,D,r) is, by definition, the set of all unitary flat D-vector bundles of rank r.

<u>NOTE 3.3.9</u> Professor Maruyama has indicated that Mehta-Ramanathan [40] recently proved the following theorem:

<u>THEOREM</u> <u>(Mehra-Ramanathan)</u> Let M be an n-dimensional projective manifold. Let H be an ample line bundle on M. Let V be a vector bundle on M with $c_1(V) = 0$ and $c_2(V).H^{n-2} = 0$. Then V comes from an irreducible unitary representation of the fundamental group $\pi_1(M)$ if and only if V is stable (with respect to H).

3.4 Tannaka systems

We recall here some facts on the *Tannaka duality*, (see Tannaka [60], [61] or Chevalley [5]).

In this section, let G be a topological group.

We denote by Uni(G,r) the set of all continuous unitary representations of G of degree r. Put

$$\mathrm{Uni}(G) = \bigcup_{r=1}^{\infty} \mathrm{Uni}(G,r) \quad \text{(the disjoint union)}$$

and

$$\mathrm{Uni}(G)/\approx \; = \bigcup_{r=1}^{\infty} \mathrm{Uni}(G,r)/\approx .$$

Under the direct sum

$$(R_1/\approx) \oplus (R_1/\approx) = \begin{pmatrix} R_1 & 0 \\ 0 & R_2 \end{pmatrix} / \approx,$$

the tensor product

$$(R_1/\approx) \otimes (R_2/\approx) = (R_1 \otimes R_2)/\approx$$

and the $*$-operation (an involution)

$$(R/\approx)^* = {}^tR^{-1}/\approx \; (= \bar{R}/\approx),$$

the set Uni(G)/≈ becomes an associative, distributive, commutative, symmetric algebraic system with identity 1 (= 1/≈). (The $*$-operation is an involutive automorphism with respect to $\oplus$ and $\otimes$.) The degree deg(R/≈) = deg(R) defines a homomorphism

$$\deg : \mathrm{Uni}(G)/\approx \longrightarrow \mathbb{N} \quad (= \text{the set of positive integers})$$

between the algebraic systems such that deg(1) = 1 and deg((R/≈)*) = deg(R/≈).

In general, an associative, distributive, commutative, symmetric algebraic system T with respect to +, x, and *-operation, without 0, but with the identity 1 and with a homomorphism

$$\deg : T \longrightarrow \mathbb{N}$$

such that $\deg(1) = 1$ and $\deg(x^*) = \deg(x)$ for $x \in T$, is called a *Tannaka system.*

$\mathbb{N}$ is itself clearly an example of a Tannaka system.

We define a *subsystem of a Tannaka system* T as a subset S of T as a subset S of T with $1 \in S$ such that S is closed under +, x and *. Then S is itself a Tannaka system.

For Tannaka systems T_1 and T_2, a map $\Phi : T_1 \longrightarrow T_2$ is called a *homomorphism* if

$$\Phi(x+y) = \Phi(x) + \Phi(y), \quad \Phi(xy) = \Phi(x)\Phi(y), \quad \Phi(1) = 1,$$

$$\Phi(x^*) = \Phi(x)^* \quad \text{and} \quad \deg\Phi(x) = \deg(x)$$

for x, y $\in$ T. Thus we have a *category of Tannaka systems.*

Next, consider the disjoint union

$$\mathbb{U} = \bigcup_{r=1}^{\infty} U(r)$$

of unitary groups. For a Tannaka system T, consider three mappings

$$f_1 : x \in T \longmapsto A_x \in \mathbb{U}$$

$$f_2 : (x,y) \in T \times T \longmapsto B_{x,y} \in \mathbb{U}$$

$$f_3 : (x,y) \in T \times T \longmapsto C_{x,y} \in \mathbb{U}$$

such that

$$\deg(A_x) = \deg(x),$$

$$\deg(B_{x,y}) = \deg(x) + \deg(y),$$

$$\deg(C_{x,y}) = \deg(x)\deg(y)$$

for $x,y \in T$. We put $\Lambda = \{f_1,f_2,f_3\}$ and call it a *transition set of* T.

<u>DEFINITION 3.4.1</u> Let $\Lambda = \{f_1,f_2,f_3\}$ be a transition set of T. A map $\Psi : T \longrightarrow \mathbb{U}$ is called a *representation of* T *with respect to* Λ if the following conditions are satisfied:

$$\Psi(1) = 1, \qquad \deg\Psi(x) = \deg(x),$$

$$\Psi(x^*) = A_x\, \overline{\Psi(x)} A_x^{-1},$$

$$\Psi(x+y) = B_{x,y} \begin{pmatrix} \Psi(x) & 0 \\ 0 & \Psi(y) \end{pmatrix} B_{x,y}^{-1},$$

$$\Psi(xy) = C_{x,y}(\Psi(x) \otimes \Psi(y))C_{x,y}^{-1}$$

for any $x,y \in T$.

The set $G_\Lambda(T)$ of all representations of T with respect to Λ forms a group be defining

$$(\Psi_1\Psi_2)(x) = \Psi_1(x)\Psi_2(x) \quad (\in U(\deg(x)),$$

$$1(x) = \begin{pmatrix} 1 & & 0 \\ & \ddots & \\ 0 & & 1 \end{pmatrix} \quad (\in U(\deg(x)).$$

We introduce a topology on $G_\Lambda(T)$ as follows: Let Ψ_o be an element of $G_\Lambda(T)$. A complete open neighbourhood system of Ψ_o is given by the sets

$$W(\Psi_o;x_1,\ldots,x_m;\varepsilon) = \{\Psi \,|\, \|\Psi(x_j) - \Psi_o(x_j)\| < \varepsilon \quad \text{for } 1 \leq j \leq m\},$$

where $\varepsilon > 0$ and $x_1,\ldots,x_m$ are any finite number of points of T. ($\|A\| = \sqrt{\mathrm{Trace}({}^t\bar{A}A)}$.) Then it is easy to see that $G_\Lambda(T)$ is a topological group.

<u>LEMMA 3.4.2</u> $G_\Lambda(T)$ is a compact topological group.

<u>PROOF</u>. Consider the compact topological group

$$\mathbb{G} = \prod_{x\in T} U(\deg(x)).$$

Consider the homomorphism

$$i : \Psi \in G_\Lambda(T) \longmapsto (\ldots,\Psi(x),\ldots) \in \mathbb{G}.$$

This is clearly an injective continuous homomorphism. The image $i(G_\Lambda(T))$ is closed in $\mathbb{G}$, hence is compact. (The limit in $\mathbb{G}$ of elements of $G_\Lambda(T)$ is also a representation of T with respect to Λ.) By the definition of the topologies of $G_\Lambda(T)$ and $\mathbb{G}$, it is clear that $i : G_\Lambda(T) \longrightarrow i(G_\Lambda(T))$ is an isomorphism between topological groups. □

For the Tannaka system $\mathrm{Uni}(G)/\approx$, there is a special choice of transition sets. To see this, we first prove

LEMMA 3.4.3 If two unitary representations R and R' of G are equivalent, then there is a unitary matrix L such that $R'(\varphi) = LR(\varphi)L^{-1}$ for all $\varphi \in G$.

PROOF. By the assumption, there is a non-singular matrix C such that

$$R'(\varphi) = CR(\varphi)C^{-1} \quad \text{for all } \varphi \in G.$$

We have then

$$R(\varphi)^{-1}\,{}^t\overline{C} = {}^t\overline{R(\varphi)}\,{}^t\overline{C} = {}^t\overline{(CR(\varphi))} = {}^t\overline{(R'(\varphi)C)} = {}^t\overline{C}\,{}^t\overline{R'(\varphi)}$$
$$= {}^t\overline{C}R'(\varphi)^{-1} = {}^t\overline{C}CR(\varphi)^{-1}C^{-1}.$$

Hence

$$({}^t\overline{C}C)R(\varphi) = R(\varphi)({}^t\overline{C}C) \quad \text{for all } \varphi \in G.$$

Note that ${}^t\overline{C}C$ is a positive definite Hermitian matrix.

Suppose first that the representation R is irreducible. Then, by Schur's lemma,

$${}^tCC = \lambda E_r,$$

where λ is a positive number and E_r is the identity matrix of degree r = the degree of R. Put

$$L = C/\sqrt{\lambda}.$$

Then L is a unitary matrix which satisfies

$$R'(\varphi) = LR(\varphi)L^{-1} \quad \text{for all } \varphi \in G.$$

Next suppose that R is reducible. By the complete reducibility of R, R can be written as

$$R(\varphi) = \begin{pmatrix} R_1(\varphi) & & 0 \\ & \ddots & \\ 0 & & R_2(\varphi) \end{pmatrix} \quad \text{for } \varphi \in G,$$

where $R_1, \ldots, R_t$ are irreducible unitary representations of degree $r_1, \ldots, r_t$, respectively, $(r_1 + \ldots + r_t = r)$.

For simplicity, we consider the case $t = 2$:

$$R(\varphi) = \begin{pmatrix} R_1(\varphi) & 0 \\ 0 & R_2(\varphi) \end{pmatrix} \quad \text{for } \varphi \in G,$$

(The general case can be treated similarly.) We write

$$^t\overline{C}C = \begin{pmatrix} A_{11} & A_{12} \\ A_{21} & A_{22} \end{pmatrix},$$

where A_{jk} is an $(r_j \times r_k)$-matrix for $1 \leq j,k \leq 2$. Then the relation

$$\begin{pmatrix} A_{11} & A_{12} \\ A_{21} & A_{22} \end{pmatrix} \begin{pmatrix} R_1(\varphi) & 0 \\ 0 & R_2(\varphi) \end{pmatrix} = \begin{pmatrix} R_1(\varphi) & 0 \\ 0 & R_2(\varphi) \end{pmatrix} \begin{pmatrix} A_{11} & A_{12} \\ A_{21} & A_{22} \end{pmatrix}$$

means that

$$A_{11}R_1(\varphi) = R_1(\varphi)A_{11}, \qquad A_{12}R_2(\varphi) = R_1(\varphi)A_{12}$$

$$A_{21}R_1(\varphi) = R_2(\varphi)A_{21}, \qquad A_{22}R_2(\varphi) = R_2(\varphi)A_{22}$$

for $\varphi \in G$. Hence, by Schur's lemma,

$$A_{11} = \lambda_{11}E_{r_1} \qquad \text{and} \qquad A_{22} = \lambda_{22}E_{r_2}$$

for some $\lambda_{11}, \lambda_{22} \in \mathbb{C}$.

If R_1 and R_2 are not equivalent, then $A_{12} = 0$ and $A_{21} = 0$ by Schur's lemma. In this case,

$$^t\overline{C}C = \begin{pmatrix} \lambda_{11}E_{r_1} & 0 \\ 0 & \lambda_{22}E_{r_2} \end{pmatrix}.$$

Hence λ_{11} and λ_{22} are positive numbers. Put

$$L = C \begin{pmatrix} E_{r_1}/\sqrt{\lambda_{11}} & 0 \\ 0 & E_{r_2}/\sqrt{\lambda_{22}} \end{pmatrix}.$$

Then L is a unitary matrix which satisfies

$$R'(\varphi) = LR(\varphi)L^{-1} \qquad \text{for all} \qquad \varphi \in G.$$

If R_1 and R_2 are equivalent, then $r_1 = r_2 = r/2$ $(= r')$ and

$$A_{12} = \lambda_{12}E_{r'}, \qquad \text{and} \qquad A_{21} = \lambda_{21}E_{r'}$$

for some $\lambda_{12}, \lambda_{21} \in \mathbb{C}$. Note that

$$\begin{pmatrix} A_{11} & A_{12} \\ A_{21} & A_{22} \end{pmatrix} = \begin{pmatrix} \lambda_{11}E_{r'} & \lambda_{12}E_{r'} \\ \lambda_{21}E_{r'} & \lambda_{22}E_{r'} \end{pmatrix}$$

is a positive definite Hermitian matrix if and only if

$$\begin{pmatrix} \lambda_{11} & \lambda_{12} \\ \lambda_{21} & \lambda_{22} \end{pmatrix}$$

is so. Hence $\lambda_{11}, \lambda_{22} \in \mathbb{R}$ and $\lambda_{21} = \overline{\lambda_{12}}$. There is then a non-singular matrix

$$\begin{pmatrix} \mu_{11} & \mu_{12} \\ \mu_{21} & \mu_{22} \end{pmatrix}$$

such that

$$\begin{pmatrix} \lambda_{11} & \lambda_{12} \\ \lambda_{21} & \lambda_{22} \end{pmatrix} = {}^t\overline{\begin{pmatrix} \mu_{11} & \mu_{12} \\ \mu_{21} & \mu_{22} \end{pmatrix}} \begin{pmatrix} \mu_{11} & \mu_{12} \\ \mu_{21} & \mu_{22} \end{pmatrix}.$$

Since R_1 and R_2 are equivalent irreducible unitary representations, there is a unitary matrix A'_o of degree r' such that

$$R_2(\varphi) = A'_o R_1(\varphi) A'^{-1}_o \qquad \text{for all } \varphi \in G.$$

Put

$$A_o = \begin{pmatrix} E_{r'} & 0 \\ 0 & A'_o \end{pmatrix}$$

Then A_o is a unitary matrix of degree r. Put

$$L = C \begin{pmatrix} \mu_{11}E_{r'} & \mu_{12}E_{r'} \\ \mu_{21}E_{r'} & \mu_{22}E_{r'} \end{pmatrix}^{-1} A_o^{-1}.$$

Then L is a unitary matrix which satisfies

$$R'(\varphi) = LR(\varphi)L^{-1} \qquad \text{for all } \varphi \in G. \qquad \square$$

Now, for the Tannaka system $Uni(G)/\approx$, we choose a unitary representation R_x from every equivalence class $x \in Uni(G)/\approx$. That is, we consider a map

$$f_o : x \in Uni(G)/\approx \longmapsto R_x \in Uni(G) \qquad \ldots(1)$$

such that $R_x/\approx = x$. We fix f_o.

By Lemma 3.4.3, there are unitary matrices $A_x, B_{x,y}$ and $C_{x,y}$ for any $x,y \in Uni(G)/\approx$ such that

$$\left.\begin{aligned} R_{x*}(\varphi) &= A_x\overline{R_x(\varphi)}A_x^{-1}, \\ R_{x+y}(\varphi) &= B_{x,y}\begin{pmatrix} R_x(\varphi) & 0 \\ 0 & R_y(\varphi) \end{pmatrix}B_{x,y}^{-1}, \\ R_{xy}(\varphi) &= C_{x,y}(R_x(\varphi) \otimes R_y(\varphi))C_{x,y}^{-1} \end{aligned}\right\} \qquad \ldots(2)$$

for all $\varphi \in G$. Consider the transition set

$$\Lambda_o = \Lambda(f_o, f_1, f_2, f_3) = \{f_1, f_2, f_3\} \qquad \ldots(3)$$

where

$$f_1 : x \longmapsto A_x$$

$$f_2 : (x,y) \longmapsto B_{x,y}$$

$$f_3 : (x,y) \longmapsto C_{x,y}$$

We fix Λ_o.

By abuse of notation, we denote $G_{\Lambda_o}(Uni(G)/\approx)$ by $G(Uni(G)/\approx)$ and call it (resp. an element of it) the *representation group* (resp. *representation*) of $Uni(G)/\approx$.

<u>REMARK 3.4.4</u> If we take another map

$$f'_o : x \in Uni(G)/\approx \longmapsto R'_x \in Uni(G),$$

such that $R'_x/\approx = x$, then, for every $x \in Uni(G)/\approx$, there is a unitary matrix

L_x with $\deg(L_x) = \deg(x)$ such that

$$R'_x(\varphi) = L_x R_x(\varphi) L_x^{-1} \quad \text{for all } \varphi \in G.$$

We may then choose the unitary matrices A'_x, $B'_{x,y}$ and $C'_{x,y}$ corresponding to f'_o as follows:

$$A'_x = L_{x*} A_x L_x^{-1},$$

$$B'_{x,y} = L_{x+y} B_{x,y} \begin{pmatrix} L_x^{-1} & 0 \\ 0 & L_y^{-1} \end{pmatrix},$$

$$C'_{x,y} = L_{xy} C_{x,y} (L_x^{-1} \otimes L_y^{-1}).$$

If Ψ is a representation of $\mathrm{Uni}(G)/\approx$ with respect to Λ_o, then Ψ' defined by

$$\Psi'(x) = L_x \Psi(x) L_x^{-1}$$

is a representation of $\mathrm{Uni}(G)/\approx$ with respect to $\Lambda'_o = \Lambda(f'_o, f'_1, f'_2, f'_3)$ where f'_1, f'_2 and f'_3 are defined by A'_x, $B'_{x,y}$ and $C'_{x,y}$ respectively. It can be then easily shown that

$$\Psi \in G(\mathrm{Uni}(G)/\approx) = G_{\Lambda_o}(\mathrm{Uni}(G)/\approx) \longmapsto \Psi' \in G_{\Lambda'_o}(\mathrm{Uni}(G)/\approx)$$

is an isomorphism between topological groups.

Now, from the relations (2) above, every $\varphi \in G$ defines a representation

$$\Psi_\varphi : x \in \mathrm{Uni}(G)/\approx \longmapsto R_x(\varphi) \in U(\deg(x))$$

of $\mathrm{Uni}(G)/\approx$. It is then easy to see that the map

$$j : \varphi \in G \longmapsto \Psi_\varphi \in G(\mathrm{Uni}(G)/\approx)$$

is a continuous homomorphism. The beautiful theorem of Tannaka [60] says

THEOREM 3.4.5 (Tannaka duality theorem) If G is compact, then $j : G \longrightarrow G(\mathrm{Uni}(G)/\approx)$ is an isomorphism between topological groups G and $G(\mathrm{Uni}(G)/\approx)$.

Now, let T be a Tannaka system. An element $x \in T$ is said to be *reducible* if x can be written as

$$x = y + z \quad \text{for some } y, z \in T.$$

Otherwise, x is said to be *irreducible*. If $x = y + z$, then $\deg(x) = \deg(y) + \deg(z)$. Hence every element x of T can be written as a sum of irreducible elements:

$$x = x_1 + \ldots + x_t.$$

However this expression may not be unique. If, for every $x \in T$, the expression

$$x = x_1 + \ldots + x_t \qquad (x_j\text{: irreducible})$$

is unique, then T is called a *unique factorization Tannaka system.* In this case, each x_j is called an *irreducible component of* x and $x = x_1 + \ldots + x_t$ is called the *irreducible decomposition of* x.

It is known that $\mathrm{Uni}(G)/\approx$ is a unique factorization Tannaka system, (see, for example, Chevalley [5, p.175]).

Suppose that a unique factorization Tannaka system T is given. A sub-system S of T is called a *module of* T if, for any $x \in S$, every irreducible component (in T) of x also belongs to S.

T is itself a module of T. If $\{S_\alpha\}$ is a set of modules of T, then $\bigcap_\alpha S_\alpha$ is also a module of T. For a subset Z of T, we denote by S_Z the smallest module which contains Z.

DEFINITION 3.4.6 A subset Z of $\mathrm{Uni}(G)/\approx$ is said to *contain sufficiently many representation classes* if, for any distinct elements φ_1 and φ_2 of G, there is $x \in Z$ such that $R'_x(\varphi_1) \neq R'_x(\varphi_2)$, where $x \in Z \longmapsto R'_x \in \mathrm{Uni}(G)$ is a map such that $R'_x/\approx \; = x$.

Note that this definition does not depend on the choice of the map $x \in Z \longmapsto R'_x \in \mathrm{Uni}(G)$.

We now state the important theorem of Kampen [36] without proof.

THEOREM 3.4.7 (Kampen) Let G be compact. If $Z \subset \mathrm{Uni}(G)/\approx$ is a subset which contains sufficiently many representation classes, then $S_Z = \mathrm{Uni}(G)/\approx$.

Let T be a unique factorization Tannaka system and Λ be a transition set of T. Take a module S of T. Λ is then a transition set of S. We apply Kampen's theorem to the compact group $G_\Lambda(S)$. By the definition of $G_\Lambda(S)$, every element $x \in S$ defines a continuous unitary representation

$$A_x : \Psi \in G_\Lambda(S) \longmapsto \Psi(x) \in U(\deg(x))$$

of $G_\Lambda(S)$. If Ψ_1 and Ψ_2 are distinct elements of $G_\Lambda(S)$, then there must exist a point $x \in S$ such that $\Psi_1(x) \neq \Psi_2(x)$. This means that $A_x(\Psi_1) \neq A_x(\Psi_2)$. Hence

$$Z_S = \{A_x/\approx \mid x \in S\}$$

is a subset of $\mathrm{Uni}(G_\Lambda(S))/\approx$ which contains sufficiently many representation classes. Thus by Kampen's theorem,

<u>PROPOSITION 3.4.8</u> Under the above notations,

$$S_{Z_S} = \mathrm{Uni}(G_\Lambda(S))/\approx.$$

<u>DEFINITION 3.4.9</u> A module S of T is said to be *finite with respect to* Λ if (1) there is a finite subset Z of S such that $S = S_Z$ and (2) $G_\Lambda(S)$ is a finite group.

We may assume that the finite set Z in the definition consists of irreducible elements of T and that $Z^* = Z$, that is, Z is closed under the *-operation.

Now, consider the special case:

$$T = \mathrm{Uni}(G)/\approx \quad \text{and} \quad \Lambda = \Lambda_o,$$

where Λ_o is the fixed transition set defined in (3) above.

Let S be a module of $\mathrm{Uni}(G)/\approx$. Put

$$N(S) = \bigcap_{x \in S} \ker(R_x) = \{\varphi \in G \mid R_x(\varphi) = E_r \quad \text{for all} \quad x \in S\}.$$

($r = \deg(x)$.) Then $N(S)$ is a closed normal subgroup of G.

Note that $N(S)$ does not depend on f_o in (1) above, but depends only on S.

We have defined a continuous homomorphism

$$j : \varphi \in G \longmapsto \Psi_\varphi \in G(\mathrm{Uni}(G)/\approx),$$

where

$$\Psi_\varphi(x) = R_x(\varphi) \quad \text{for} \quad x \in \mathrm{Uni}(G)/\approx.$$

By abuse of notation, we denote the restriction $\Psi_\lambda|S$ of Ψ_φ to S by Ψ_φ again. This is a representation of S with respect to Λ_o. Also, we put

$$G(S) = G_{\Lambda_o}(S).$$

We have thus a continuous homomorphism

$$j_S : \varphi \in G \longrightarrow \Psi_\varphi \in G(S).$$

We clearly have

LEMMA 3.4.10 $N(S) = \ker(j_S)$.

A module S is said to be *finite* if it is finite with respect to Λ_o.

LEMMA 3.4.11 If S is a finite module of $\mathrm{Uni}(G)/\approx$, then (1) S is isomorphic to $\mathrm{Uni}(G/N(S))/\approx$ (as Tannaka systems) and (2) $G(S)$ is isomorphic to $G/N(S)$. In particular, $N(S)$ is of finite index in G.

PROOF. Let $Z = \{x_1,\dots,x_t\}$ be a finite set of irreducible elements of $\mathrm{Uni}(G)/\approx$ such that $Z^* = Z$ and $S = S_Z$. Put

$$x_o = x_1 + \dots + x_t.$$

Then it is clear that

$$\ker(R_{x_o}) = \bigcap_{j=1}^{t} \ker(R_{x_j}).$$

It is also clear that, for any $x \in S$,

$$\ker(R_{x_o}) \subset \ker(R_x).$$

This shows that

$$N(S) = \ker(R_{x_o}).$$

For every $x \in S$, R_x splits as follows:

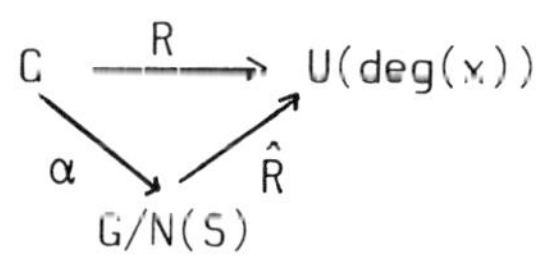

where α is the canonical projection. Then

$$\hat{S} = \{\hat{x} = \hat{R}_x/\approx \mid x \in S\}$$

is a module of $\mathrm{Uni}(T/N(S))/\approx$. The correspondence

$$x \in S \longmapsto \hat{x} = \hat{R}_x/\approx \in \hat{S}$$

is then an isomorphism (as for Tannaka systems). Since

$$\hat{R}_{x_o} : G/N(S) \longrightarrow U(\deg(x_o))$$

is injective, $\hat{S}$ contains sufficiently many representation classes.

On the other hand, by Lemma 3.4.10, $j_S : G \longrightarrow G(S)$ induces an injective continuous homomorphism

$$\hat{j}_S : G/N(S) \longrightarrow G(S).$$

By the assumption G(S) is a finite group. Hence G/N(S) is also a finite group.

Thus we may apply Kampen's theorem (Theorem 3.4.7) to $\hat{S}$:

$$\hat{\hat{S}} = S_{\hat{S}} = \mathrm{Uni}(G/N(S))/\approx.$$

Hence

$$S \simeq \mathrm{Uni}(G/N(S))/\approx \qquad \text{(as for Tannaka systems).}$$

By the Tannaka duality theorem (Theorem 3.4.5),

$$G(S) \simeq G(\hat{S}) \simeq G/N(S). \quad \square$$

Finally, we prove

THEOREM 3.4.12 Let G be a topological group. The correspondence

$$S \longmapsto N = N(S)$$

$$(S = S(N) \longleftarrow N)$$

is a one-to-one correspondence between the set of all finite modules S of Uni(G)/≈ and the set of all closed normal subgroups N of G of finite index, such that

$$G/N \simeq G(S) \quad \text{and} \quad S \simeq \mathrm{Uni}(G/N)/\approx.$$

PROOF. Let N be a closed normal subgroup of G of finite index. Put

$$\hat{S} = \mathrm{Uni}(G/N)/\approx.$$

Consider a map

$$\hat{f}_o : \hat{x} \in \hat{S} \longmapsto \hat{R}_{\hat{x}} \in \mathrm{Uni}(G/N)$$

such that $\hat{R}_{\hat{x}}/\approx\, = \hat{x}$. Put

$$S = S(N) = \{x = (\hat{R}_{\hat{x}}.\alpha)/\approx \mid \hat{x} \in \hat{S}\},$$

where $\alpha : G \longrightarrow G/N$ is the canonical projection. Then S is a module of Uni(G)/≈ such that

$$x \in \hat{S} \longmapsto x = (\hat{R}_{\hat{x}}.\alpha)/\approx\, \in S$$

is an isomorphism as for Tannaka systems.

Note that S does not depend on the choice of $\hat{f}_o$.

By the Tannaka duality theorem (Theroem 3.4.5),

$$G(S) \simeq G(\hat{S}) \simeq G/N.$$

Hence G(S) is a finite group.

On the other hand, there is a representation $\hat{R}_0 \in \mathrm{Uni}(G/N)$ such that $\hat{R}_0 : G/N \longrightarrow U(\deg(\hat{R}_0))$ is injective. (For example, let $\hat{R}_0$ be a unitary representation which is equivalent to the regular representation of G/N.) Put $\hat{x}_0 = \hat{R}_0/\approx$. Then, by the representation theory of finite groups,

$$\hat{S} = S_{\{\hat{x}_0\}}.$$

Hence

$$S = S_{\{x_0\}},$$

where $x_0 = (\hat{R}_0.\alpha)/\approx$. Thus S is a finite module of $\mathrm{Uni}(G)/\approx$. It is clear that

$$N = N(S) \quad (= \ker(\hat{R}_0.\alpha))$$

under the previous notation. □

NOTE 3.4.13 For a more abstract treatment of the content of this section, one can use the *Tannaka category*. See Saavedva Rivano [53], Deligne-Milne [10].

3.5 Galois extensions

Again, we follow the notations and the assumptions in section 3.1. (But the main results of this section turn out to be free from Condition 1.2.4 as will be shown.) Under the discrete topology, we regard $\tilde{G}$ as a topological group. Consider the unique factorization Tannaka system

$$\mathrm{Uni}(\tilde{G})/\approx \; = \bigcup_{r=1}^{\infty} \mathrm{Uni}(\tilde{G},r)/\approx$$

of all equivalence classes of unitary representations of $\tilde{G}$.

EXAMPLE 3.5.1 Let $\bar{M} = M$ be a compact Riemann surface of genus g and $\bar{D} = D = e_1p_1 + \ldots + e_sp_s$. Then, under the notations in section 1.2,

$$\tilde{G} = \langle \alpha_1,\beta_1,\ldots,\alpha_g,\beta_g,\gamma_1,\ldots,\gamma_s \mid \gamma_1^{e_1} = \ldots = \gamma_s^{e_s} = 1 \text{ and } \alpha_1\beta_1\alpha_1^{-1}\beta_1^{-1}\ldots\alpha_g\beta_g\alpha_g^{-1}\beta_g^{-1}\gamma_1\ldots\gamma_s = 1\rangle.$$

Hence $\mathrm{Uni}(\tilde{G})/\approx$ can be identified with the Tannaka system

$$\bigcup_{r=1}^{\infty} W_r/U(r)$$

with the operations x, +, and * defined naturally, where

$$W_r = \{(A_1,B_1,\dots,A_g,B_g,C_1,\dots,C_s) \in U(r)^{2g+s} \mid C_1^{e_1} = \dots$$

$$= C_s^{e_s} = 1,\ A_1B_1A_1^{-1}B_1^{-1}\dots A_gB_gA_g^{-1}B_g^{-1}C_1\dots C_s = 1\},$$

and the action of $U(r)$ on W_r is defined by

$$(A_1,B_1,\dots,A_g,B_g,C_1,\dots,C_s)$$

$$\longmapsto (QA_1Q^{-1},QB_1,Q^{-1},\dots,QA_gQ^{-1},QB_gQ^{-1},QC_1Q^{-1},\dots,QC_sQ^{-1})$$

for $Q \in U(r)$.

By the discussion of section 3.3, there are the following bijective correspondences:

$$\mathrm{Uni}(\tilde{G})/\approx \longrightarrow \mathbf{M}_s(M,D)/\approx \longrightarrow \mathrm{Div}_s(M,D)/\sim$$

$$\longrightarrow \mathrm{UFV}(M,D),$$

where

$$\mathbf{M}_s(M,D) = \bigcup_{r=1}^{\infty} M_s(M,D,r),$$

$$\mathrm{Div}_s(M,D) = \bigcup_{r=1}^{\infty} \mathrm{Div}_s(M,D,r),$$

and

$$\mathrm{UFV}(M,D) = \bigcup_{r=1}^{\infty} \mathrm{UFV}(M,D,r).$$

Here $\mathrm{UFV}(M,D,r)$ is the set of all (isomorphism classes of) unitary flat D-vector bundles of degree r.

A unitary flat D-vector bundle V is said to be *reducible* if $V = V_1 \oplus V_2$ for some unitary flat D-vector bundles V_1 and V_2. Otherwise, V is said to be *irreducible*. If V is indecomposable (see section 3.2), then V is irreducible. (We do not know if the converse holds.)

Note that $\mathrm{UFV}(M,D)$ is a unique factorization Tannaka system under the direct sum, tensor product and the *-operation

$$* : V \longrightarrow V^* = {}^tV^{-1}$$

of unitary flat D-vector bundles. Moreover the above bijection

$$\varkappa : \mathrm{Uni}(\tilde{G})/\approx \longrightarrow \mathrm{UFV}(M,D)$$

is an isomorphism between the Tannaka systems.

We may introduce the structures of Tannaka systems on both $\mathbf{M}_s(M,D)/\approx$ and $\mathrm{Div}_s(M,D)/\sim$ through the above bijections, but we do not use them directly in the following discussion.

Let Λ_o be the transition set of UFV(M,D) which corresponds to the fixed Λ_o of $\mathrm{Uni}(\tilde{G})/\approx$ in section 3.4, through the isomorphism $\varkappa$. (Λ_o can be expressed in terms of the period representation in Theorem 3.3.6.) We denote $G_{\Lambda_o}(S)$ simply by $G(S)$ for any module S of UFV(M,D).

The following theorem can be considered as a generalization of Theorem 2.3.17 to Galois extensions:

THEOREM 3.5.2 There is a one-to-one correspondence between the set of all finite Galois extensions K of $\mathbb{C}(M)$ with $K \subset \mathbb{C}(\tilde{M})$ and the set of all finite modules S of UFV(M,D). The correspondence is given by

$$K \longmapsto S = S(K) = S_Z$$

(S_Z = the smallest module which contains Z), where

$$Z = \{[(F)] \mid F = (f_{jk}) \text{ is a strict D-multiplicative matrix function such that } f_{jk} \in K \text{ for all } j,k\}.$$

Moreover, the correspondence satisfies (1) $\mathrm{Gal}(K/\mathbb{C}(\tilde{M})) \simeq G(S(K))$ and (2) $K_1 \subset K_2$ if and only if $S(K_1) \subset S(K_2)$.

PROOF. By Theorem 2.1.3 and Theorem 3.4.12, we have the bijective correspondences

$$K \longmapsto N = G(K) \longmapsto S(N) \overset{\varkappa}{\longmapsto} S(K).$$

Under the notations in the proof of Theorem 3.4.12,

$$S(N) = S_{\{x_o\}},\ x_o = R_o/\approx,\quad R_o = \hat{R}_o.\alpha$$

and

$$N = \ker(R_o).$$

Let $x_o = x_1 + \dots + x_t$ be the irreducible decomposition of x_o. Then there is a unitary matrix C such that

$$R_o = C\begin{pmatrix} R_1 & & \\ & \ddots & \\ & & R_t \end{pmatrix}C^{-1}$$

where $R_\nu = R_{x_\nu}$ for $1 \leq \nu \leq t$. Note that

$$S(K) = S_{Z_o}$$

where

$$Z_o = \{[(F_\nu)] \mid F_\nu \text{ is a representative matrix function with respect to } R_\nu,\ 1 \leq \nu \leq t\}.$$

Note also that

$$F = C\begin{pmatrix} F_1 & & \\ & \ddots & \\ & & F_t \end{pmatrix}$$

is a representative matrix function with respect to R_o.

By Proposition 2.2.9,

$$K = K(N) = K_{R_o}\ .$$

Hence every component of F_ν belongs to K.

We define the set Z as in the statement of the theorem. Then $Z_o \subset Z$. We show that $Z \subset S(K)$. Take $[(F)] \in Z$. Let $x = R/\approx$ be the point of $\mathrm{Uni}(\tilde{G})/\approx$ which corresponds to $[(F)]$ under the isomorphism $\varkappa$. Then

$$K_R \subset K = K_{R_o}\ .$$

Hence $K_R/\mathbb{C}(M)$ is a finite extension. By Proposition 2.2.9,

$$G(K_R) = \ker(R) \supset G(K) = N.$$

Hence R splits as

$$\begin{array}{ccc} \tilde{G} & \xrightarrow{\ R\ } & U(\deg(R)) \\ & \alpha\searrow \quad \nearrow\hat{R} & \\ & \tilde{G}/N & \end{array}$$

Put $\hat{x} = \hat{R}/\approx\ \in \mathrm{Uni}(\tilde{G}/N)/\approx$. Since $\{\hat{x}_o\}$ $(\hat{x}_o = \hat{R}_o/\approx)$ contains sufficiently many representation classes, $\mathrm{Uni}(\tilde{G}/N)/\approx\ = S_{\{\hat{x}_o\}}$ by Theroem 3.4.7. This shows that $x \in S_{\{x_o\}} = S(N)$. Hence $[(F)] \in S(K)$. Thus we have $Z \subset S(K)$.

Hence

$$S(K) = S_{Z_o} = S_Z.$$

Moreover,

$$\mathrm{Gal}(K/\mathbb{C}(M)) \simeq \tilde{G}/N \simeq G(S(K)).$$

The assertion (2) of the theorem is clear, since $S(K) = S_Z$. □

The above theorem can be rewritten as follows (cf. Theorem 2.3.18):

THEOREM 3.5.3 Let $\bar{D}_j$ $(1 \le j \le s)$ be distinct irreducible hypersurfaces of a projective manifold $\bar{M}$. Let e_j $(1 \le j \le s)$ be positive integers and put $\bar{D} = e_1\bar{D}_1 + \ldots + e_s\bar{D}_s$. Put $\bar{B} = \cup\bar{D}_j$, $M = \bar{M}\text{-Sing}\bar{B}$, $D_j = \bar{D}_j\text{-Sing}\bar{B}$ $(1 \le j \le s)$ and $D = e_1D_1 + \ldots + e_sD_s$. Then there is a one-to-one correspondence between the set of all isomorphism classes of finite Galois coverings $\bar{\pi} : \bar{X} \longrightarrow \bar{M}$ which branches at most at $\bar{D}$, and the set of all finite modules S of $UFV(M,D)$. The correspondence $\bar{\pi} \longmapsto S = S(\bar{\pi})$ satisfies
(1) $G_{\bar{\pi}} \simeq G(S(\bar{\pi}))$ and (2) $\bar{\pi}_1 \le \bar{\pi}_2$ if and only if $S(\bar{\pi}_1) \subset S(\bar{\pi}_2)$.

PROOF. By Theorem 2.1.3 and Theroem 1.2.7, there is a one-to-one correspondence between the set of all finite Galois extensions K of $\mathbb{C}(M)$ with $K \subset \mathbb{C}(\tilde{M})$ and the set of all isomorphism classes of finite Galois coverings $\pi : X \longrightarrow M$ which branches at most at D. Let $\bar{\pi} : \bar{X} \longrightarrow \bar{M}$ be the $\mathbb{C}(X)$-normalization of M. Then the correspondence $\pi \longmapsto \bar{\pi}$ is one-to-one. □

REMARK 3.5.4 (cf. Remark 2.3.19) *Theorem 3.5.3 is independent of Condition 1.2.4.* In fact, let e'_j be the smallest positive integer such that $\gamma_j^{e'_j} \in J$. Then, by Lemma 1.2.3, e'_j divides e_j and Condition 1.2.4 is satisfied for (M,D'), where

$$D' = e'_1D_1 + \ldots + e'_sD_s.$$

Hence we have a maximal covering

$$\tilde{\pi} : \tilde{M}(D') \longrightarrow M.$$

Put $\tilde{G} = G_{\tilde{\pi}}$. Any covering $\pi : X \longrightarrow M$ which branches at most at D is necessarily a covering which branches at most at D' (see Remark 2.3.19).

Hence, in order to show our assertion, it suffices to show that, for any positive integer r,

$$UFV(M,D,r) = UFV(M,D',r),$$

where $UFV(M,D,r)$ is the set of all unitary flat D-vector bundles of rank r, defined at the end of section 3.3.

It is clear that

$$\mathrm{UFV}(M,D,r) \supset \mathrm{UFV}(M,D',r).$$

Let V be a unitary flat D-vector bundle of rank r. Then there is a D-divisor $\hat{E} = \{F_\alpha\}_{\alpha \in A} \cup \{F_\nu\}_{\nu \in N}$ such that $[\hat{E}] = V$, and a meromorphic matrix 1-form η on M such that $\hat{E}\eta\hat{E} - (d\hat{E})\hat{E}^{-1}$ is holomorphic and such that the period representation R_η of $\pi_1(M-B, p_o)$ is equivalent to a unitary representation, (cf. the end of section 3.3). As at the end of section 3.3, put

$$F(p) = e\int_{p_o,\gamma}^{p} \eta .$$

Then F can be considered as a matrix valued meromorphic function on the universal covering space $(M-B)_o$ of M-B.

We show that

$$R_\eta(\gamma_j^{e_j}) = e\int_{\gamma_j^{e_j}} \eta = 1 \quad \text{for } 1 \leq j \leq s.$$

In fact, every point of B has neighbourhood W_ν with a coordinate system $(w_1,\ldots,w_n)$ such that

$$B \cap W_\nu = D_n \cap W_\nu = \{w_n = 0\} \quad \text{(say)}$$

and such that $F_\nu = F_\nu(w_1,\ldots,w_{n-1}, \sqrt[e]{w_n})$ is a matrix valued meromorphic function of $(w_1,\ldots,w_{n-1}, \sqrt[e]{w_n})$ with non-zero $\det(F_\nu)$, $(e = e_n)$. As in the proof of Theorem 3.2.8,

$$e\int_{p_\nu,\gamma_\nu}^{p} \eta = F_\nu(p_\nu)^{-1} \left(e\int_{p_\nu,\gamma_\nu}^{p} \omega\right)F_\nu(w_1,\ldots,w_{n-1}, \sqrt[e]{w_n}),$$

where $p_\nu \in W_\nu$ is a fixed point, $p = (w_1,\ldots,w_n) \in W_\nu$, γ_ν is a curve in W_ν from p_ν to p, and $\omega = F_\nu\eta F_\nu^{-1} - (dF_\nu)F_\nu^{-1}$ is holomorphic in $(w_1,\ldots,w_{n-1},\sqrt[e]{w_n})$. Here

$$e\int_{p_\nu,\gamma_\nu}^{p} \omega = H(w_1,\ldots,w_{n-1},\sqrt[e]{w_n})$$

is holomorphic in $(w_1,\ldots,w_{n-1},\sqrt[e]{w_n})$. In particular, if γ_n is a closed path in W_ν which rounds $\{w_n = 0\}$ once in a counterclockwise direction, then

$$e\int_{p_\nu,\gamma_\nu\gamma_n}^{p} \eta = F_\nu(p_\nu)^{-1}H(w_1,\ldots,w_{n-1},\zeta \sqrt[e]{w_n})F_\nu(w_1,\ldots,w_{n-1},\zeta \sqrt[e]{w_n}),$$

where $\zeta = \exp(2\pi\sqrt{-1}/e)$. Hence

$$e\int_{p_\nu,\gamma_\nu\gamma_n^e}^{p} \eta = e\int_{p_\nu,\gamma_\nu}^{p} \eta .$$

In particular

$$e\int_{p_\nu,\gamma_n^e} \eta = 1.$$

Hence we get

$$R_\eta(\gamma_j^{e_j}) = e\int_{p_o,\gamma_j^{e_j}} \eta = 1 \quad \text{for} \quad 1 \leq j \leq n.$$

Hence, for every $\delta \in \pi_1(M-B,\ p_o)$,

$$R_\eta(\delta\gamma_j^{e_j}\delta^{-1}) = 1.$$

Since $\gamma_j^{e'_j} \in J$ and J is generated by $\delta\gamma_j^{e_j}\delta^{-1}$, we have

$$R_\eta(\gamma_j^{e'_j}) = e\int_{\gamma_j^{e'_j}} \eta = 1 \quad \text{for } 1 \leq j \leq s.$$

This implies that, in W_ν,

$$e\int_{p_\nu,\gamma_\nu\gamma_n^{e'}}^{p} \eta = e\int_{p_\nu,\gamma_\nu}^{p} \eta . \qquad (e' = e'_n)$$

That is,

$$F(w_1,\ldots,w_{n-1},\ \sqrt[e]{w_n}) \equiv F(w_1,\ldots,w_{n-1},\ \zeta^{e'}\ \sqrt[e]{w_n}).$$

Hence, F is, on W_ν, a meromorphic function of $(w_1,\ldots,w_{n-1},\ \sqrt[e']{w_n})$.

Thus F can be considered as a matrix valued meromorphic function on $\tilde{M}(D')$. Since $\eta = F^{-1}dF$ is a meromorphic matrix 1-form *on* M, F is a D'-multiplicative matrix function (see Lemma 3.2.4). Clearly,

$$(F) = \hat{E}$$

and

$$(\varphi^*F)F^{-1} = R_\eta(\varphi) \quad \text{for} \quad \varphi \in \tilde{G} = \pi_1(M-B,\ p_o)/J.$$

Hence

$$V = [\hat{E}] \in UFV(M,D',r).$$

Thus

$$UFV(M,D,r) = UFV(M,D',r).$$

Thus Theorem 3.5.3 is independent of Condition 1.2.4.

From this remark, we can find, conversely, a criterion (independent of Condition 1.2.4) for the existence of a finite Galois covering $\bar{\pi} : \bar{X} \longrightarrow \bar{X}$ which branches *at* $\bar{D}$ as follows, (cf. Theorem 2.3.20):

THEOREM 3.5.5 Under the same notations as in Theorem 3.5.3, there is a finite Galois covering $\bar{\pi} : \bar{X} \longrightarrow \bar{M}$ which branches *at* $\bar{D}$ if and only if there is a finite module S of UFV(M,D) such that, for every j, $1 \leq j \leq s$, there is an element $V_j \in S$ which satisfies the following conditions: there is a D-divisor $\hat{E}_j$ with $[\hat{E}_j] = V_j$ and a meromorphic matrix 1-form η_j on M such that (1) $\hat{E}_j\eta_j\hat{E}_j^{-1} - (d\hat{E}_j)\hat{E}_j^{-1}$ is holomorphic, (2) the period representation R_{η_j} of $\pi_1(M-B, p_o)$ for η_j is equivalent to a unitary representation and (3) $R_{\eta_j}(\gamma_j)$ has the order e_j.

The proof is straightforward and is omitted.

NOTE 3.5.6 In the case when $\bar{B}$ is empty, that is, in the unramified case, Nori [47] gave a result which realtes closely to Theorem 3.5.2.

In the rest of this section, we write M, B, D_j, D, etc, instead of $\bar{M}$, $\bar{B}$, $\bar{D}_j$, $\bar{D}$, etc, for simplicity.

We move D and consider various UFV(M,D)'s on a fixed projective manifold M.

We first note that

$$UFV(M,D) \subset UFV(M,D')$$

for $D = e_1D_1 + \dots + e_sD_s$ and $D' = e_1'D_1 + \dots + e_s'D_s$ such that e_j divides e_j' for $1 \leq j \leq s$. Next, we note that, if $e_1 = 1$, say, then

$$UFV(M,D) = UFV(M,D'')$$

for $D = e_1D_1 + e_2D_2 + \dots + e_sD_s$ and $D'' = e_2D_2 + \dots + e_sD_s$, because elements of them correspond to unitary representations of $\pi_1(M-B', p_o)$, where $B' = D_2 \cup \dots \cup D_s$.

Hence the union

$$UFGV(M) = \bigcup_D UFV(M,D),$$

where $\cup$ runs over all positive divisors D on M, forms naturally a Tannaka system. We call it the *Tannaka system of all unitary flat generalized vector bundles*. We call each element of UFGV(M) a *unitary flat generalized vector bundle*.

By Theorem 3.5.3, we conclude

THEOREM 3.5.7 Let M be a projective manifold. Then there is a one-to-one correspondence $\pi \longmapsto S = S(\pi)$ between the set of all isomorphism classes of finite Galois branched coverings $\pi : X \longrightarrow M$ and the set of all finite modules S of UFGV(M). The correspondence satisfies (1) $G(S(\pi)) \simeq G_\pi$ and (2) $\pi_1 \leq \pi_2$ if and only if $S(\pi_1) \subset S(\pi_2)$.

REMARK 3.5.8 Finite abelian branched coverings correspond to finite modules S which are generated by elements of rank 1. The elements of rank 1 of S form an abelian group S_1. Theorem 2.3.30 asserts that the correspondence $\pi \mapsto S_1$ is one-to-one.

Finally, Theorem 3.5.7 can be written as

THEOREM 3.5.7' Let M be a projective manifold. Then there is a one-to-one correspondence $K \longmapsto S = S(K)$ between the set of all (isomorphism classes of) finite Galois extensions $K/\mathbb{C}(M)$ and the set of all finite modules S of UFGV(M). The correspondence satisfies (1) $G(S(K)) \simeq \mathrm{Gal}(K/\mathbb{C}(M))$ and (2) $K_1 \subset K_2$ if and only if $S(K_1) \subset S(K_2)$.

References

[1] Y. Akizuki, Theory of harmonic integrals, Iwanami, 1963, (in Japanese).

[2] M. Atiyah, Complex analytic connections in fibre bundles, Trans. Amer. Math. Soc. 85 (1957), 181-207.

[3] M. Atiyah & W. Hodge, Integrals of the second kind on an algebraic variety, Ann. Math. 62 (1955), 56-91.

[4] S. Bundgaard & J. Nielsen, On normal subgroups with finite index in F-groups, Math. Tidsskrift, B (1951), 56-58.

[5] C. Chevalley, Theory of Lie groups, Princeton Univ. Press, 1946.

[6] H. Coxeter, Factor groups of the braid group, Proc. 4-th Canadian Math. Congress, 95-122.

[7] H. Coxeter & W. Moser, Generators and relations for discrete groups, Springer-Verlag, 1980.

[8] P. Deligne, Equations différentielles à points singuliers réguliers, Springer Lec. Notes 163 (1970).

[9] P. Deligne, Le groupe fondamental du complément d'une courbe plane n'ayant que des points doubles ordinaires est abélien, Sém, Bourbaki 543, Springer Lec. Notes 842 (1981), 1-10.

[10] P. Deligne & J. Milne, Tannakian categories, Springer Lec. Notes 900 (1982), 101-228.

[11] G. de Rham, Varietes differentiables, Hermann Paris, 1960.

[12] G. de Rham & K. Kodaira, Harmonic integrals, Lec. Notes, Inst. Advanced Study, Princeton, 1950.

[13] G. Fischer, Complex analytic geometry, Springer Lec. Notes 538 (1976).

[14] R. Fox, On Fenchel's conjecture about F-groups, Math. Tidsskrift, B (1952), 61-65.

[15] R. Fox, Covering spaces with singularities, Lefschetz Symposium, Princeton Univ. Press, 1957, 243-262.

[16] R. Fox & R. Crowell, Introduction to knot theory, Ginn and Company, 1963.

[17] T. Fukui, Irregularities of branched abelian coverings over the complex projective plane, to appear.

[18] W. Fulton, On the fundamental group of the complement of a node curve, Ann. Math. 111 (1980), 407-409.

[19] H. Grauert, Ein Theorem der analytischen Garbentheorie und die Modulräume komplexer Strukturen, IHES 5 (1960).

[20] H. Grauert, Über Modifikationen und exzeptionelle analytische Mengen, Math. Ann. 146 (1962), 331-368.

[21] H. Grauert & R. Remmert, Komplexe Räume, Math. Ann. 136 (1958), 245-318.

[22] P. Griffiths & J. Harris, Principles of algebraic geometry, John Wiley & Sons, 1978.

[23] A. Grothendieck, Sur le mémoire de Weil: Généralisation des fonctions abéliennes, Sém. Bourbaki 141 (1956).

[24] A. Grothendieck & M. Raynaud, Geometrie algebrique et geometrie analytique, SGA I, Springer Lec. Notes 224 (1971), 311-343.

[25] R. Gunning, Lectures on vector bundles over Riemann surfaces, Math. Notes 6 (1967), Princeton Univ..

[26] R. Gunning, Lectures on complex analytic varieties: Finite analytic mappings, Math. Notes 14 (1974), Princeton Univ..

[27] H. Hamm & Lê Dũng Tráng, Un théorème de Zariski du type de Lefschetz, Ann. Sci. Éc. Norm. Sup. 6 (1973), 317-366.

[28] H. Hironaka, Resolution of singularities of an algebraic variety over a field of characteristic zero, I, II, Ann. Math. 79 (1964), 109-326.

[29] H. Hironaka, Bimeromorphic smoothing of complex spaces, Lec. Notes, Harvard Univ., 1971.

[30] F. Hirzebruch, Arrangements of lines and algebraic surfaces, Prog. in Math. 36 (1983), Birkhäuser, 113-140.

[31] H. Hochstadt, The functions of mathematical physics, John Wiley & Sons, 1971.

[32] T. Höfer, Ballquotienten als verzweigte Überlagerungen der projektiven Ebene, Dissertation Bonn, 1985.

[33] S. Iitaka, Algebraic geometry, Springer-Verlag, 1982.

[34] M. Ishida, The irregularities of Hirzebruch's examples of surfaces of general type with $c_1^2 = 3c_2$, Math. Ann. 262 (1983), 407-420.

[35] K. Iwasawa, Theory of algebraic functions, Iwanami, 1952, (in Japanese).

[36] E. van Kampen, Almost periodic functions and compact groups, Ann. Math. 37 (1936), 78-91.

[37] M. Kato, On uniformizations of orbifolds, Advanced Studies in Pure Math. 9 (1986), North Holland, 149-172.

[38] M. Kato, On the existence of finite principal uniformization of $\mathbb{CP}^2$ along weighted line configuraitons, Mem. Kyushu Univ. 38 (1984), 127-132.

[39] M. Kita, The Riemann-Hilbert problem and its application to analytic functions of several complex variables, Tokyo J. Math. 2 (1979), 1-27.

[40] V. Mehta & A. Ramanathan, Restriction of stable sheaves and representations of the fundamental group, to appear.

[41] J. Milnor, Singular points of complex hypersurfaces, Ann. Math. Studies 61 (1968), Princeton Univ..

[42] Y. Miyaoka, On the Chern numbers of surfaces of general type, Inv. Math. 42 (1977), 225-237.

[43] S. Nakano, On complex analytic vector bundles, J. Math. Soc. Japan 7 (1955), 1-12.

[44] M. Namba, Geometry of projective algebraic curves, Marcel Dekker, 1984.

[45] R. Narasimhan, Introduction to the theory of analytic spaces, Springer Lec. Notes 25 (1966).

[46] M.S. Narasimhan & C. Seshadri, Stable and unitary vector bundles on a compact Riemann surface, Ann. Math. 82 (1965), 540-567.

[47] M. Nori, On the representation of the fundamental group, Compos. Math. 33 (1976), 29-41.

[48] T. Oda, Vector bundles on abelian surfaces, Inv. Math. 13 (1971), 247-260.

[49] H. Pinkham, Singularités de Klein, Springer Lec. Notes 777 (1980), 1-20.

[50] L. Pontrjagin, Topological groups, Princeton Univ. Press, 1939.

[51] S. Roan, Branched covering spaces, Chinese J. Math. 7 (1979), 177-206.

[52] H. Röhrl, Das Riemann-Hilbertsche Problem der Theorie der linearen Differentialgleichungen, Math. Ann. 133 (1957), 1-25.

[53] N. Saavedva Rivano, Categories Tannakienne, Springer Lec. Notes 265 (1972).

[54] A. Selberg, On discontinuous groups in higher dimensional symmetric spaces, Contributions to Function Theory, Tata Inst., Bombay, 1960, 147-164.

[55] J. Serre, Groupes algébriques et corps de classes, Hermann Paris, 1959.

[56] C. Seshadri, Space of unitary vector bundles on a compact Riemann surface, Ann. Math. 85 (1967), 303-336.

[57] C. Siegel, Analytic functions of several complex variables, Lec. Notes, Inst. Advanced Study, Princeton, 1962.

[58] E. Spanier, Algebraic topology, McGraw-Hill, 1966.

[59] N. Steenrod, The topology of fibre bundles, Princeton Univ. Press, 1951.

[60] T. Tannaka, Über den Dualitätssatz der nichtkommutativen topologischen Gruppen, Tohoku Math. J. 45 (1938), 1-12.

[61] T. Tannaka, Theory of topolgoical groups, Iwanami, 1949, (in Japanese).

[62] H. Tōyama, Über eine nicht-Abelsche Theorie der algebraischen Funktionen, Thesis, Tohoku Univ., 1949.

[63] J. Wavrik, Deformations of banach coverings of complex manifolds, Amer. J. Math. 90 (1968), 926-960.

[64] A. Weil, Généralisation des fonctions abeliennes, J. Math. Pures et Appl. 17 (1938), 47-87.

[65] A Weil, Sur la théorie des formes differentielles attachées à une variété analytique complexe, Comm. Math. Helv. 20 (1947), 110-116.

[66] A Weil, Basic number theory, Springer-Verlag, 1967.

[67] T. Yamazaki & M. Yoshida, On Hirzebruch's examples of surfaces with $c_1^2 = 3c_2$, Math. Ann. 266 (1984), 421-431.

[68] S. T. Yau, On Calabi's conjecture and some new results in algebraic geometry, Proc. Nat. Acad. Sci. USA 74 (1977), 1798-1799.

[69] O. Zariski, Collected papers, Vol. III, 1978.

Index

34